Measuring, modeling and mitigating biodynamic feedthrough

Joost Venrooij

Bibliografische Information der Deutschen Nationalbibliothek

Die Deutsche Nationalbibliothek verzeichnet diese Publikation in der
Deutschen Nationalbibliografie; detaillierte bibliografische Daten sind
im Internet über http://dnb.d-nb.de abrufbar.

ISBN 978-3-8325-4105-7

Logos Verlag Berlin GmbH
Comeniushof, Gubener Str. 47,
10243 Berlin
Tel.: +49 (0)30 42 85 10 90
Fax: +49 (0)30 42 85 10 92
INTERNET: http://www.logos-verlag.de

Measuring, modeling and mitigating biodynamic feedthrough

Proefschrift

ter verkrijging van de graad van doctor
aan de Technische Universiteit Delft,
op gezag van de Rector Magnificus
prof. ir. K. C. A. M. Luyben,
voorzitter van het College voor Promoties,
in het openbaar verdedigd
op vrijdag 21 maart 2014 om 12.30 uur

door

Joost VENROOIJ

Ingenieur Luchtvaart en Ruimtevaart

geboren te Vught

Dit proefschrift is goedgekeurd door de promotoren:
prof. dr. ir. M. Mulder
prof. dr. H. H. Bülthoff

Copromotor: dr. ir. D. A. Abbink

Samenstelling promotiecommissie:

Rector Magnificus, voorzitter
Technische Universiteit Delft

prof. dr. ir. M. Mulder, promotor
Technische Universiteit Delft

prof. dr. H. H. Bülthoff, promotor
Max-Planck-Institut für biologische Kybernetik

dr. ir. D. A. Abbink, copromotor
Technische Universiteit Delft

prof. M. J. Griffin, B.Sc., Ph.D.
University of Southampton

prof. dr. R. Babuska
Technische Universiteit Delft

R. B. Gillespie, Ph.D.
University of Michigan

P. Masarati, Ph.D.
Politecnico di Milano

prof. dr. F. C. T. van der Helm, eerste reservelid
Technische Universiteit Delft

prof. dr. ir. J. A. Mulder, tweede reservelid
Technische Universiteit Delft

Dr. ir. Marinus M. van Paassen en dr. ir. Mark Mulder hebben als begeleiders in belangrijke mate aan de totstandkoming van het proefschrift bijgedragen.

Preface

This is the second edition of this dissertation. The first edition was published under the same title in March 2014 (ISBN: 987-94-6259-096-0). One reason for publishing a second edition of this work is that almost all copies of the first edition have been distributed. Another – and more important – reason is that new material has become available for inclusion in this edition.

The most important changes can be found in Chapter 8. This chapter has been revised and includes new material. Another important improvement in this edition is that many figures are printed in color, significantly improving readability. Finally, several corrections were made throughout the thesis.

Joost Venrooij   *Tübingen, August 2015*

Summary

Measuring, modeling and mitigating biodynamic feedthrough

Joost Venrooij

V EHICLE accelerations affect the human body in various ways. In some cases, accelerations cause involuntary motions of limbs like arms and hands. If someone is engaged in a manual control task at the same time, these involuntary limb motions can lead to involuntary control forces and control inputs. This phenomenon is called **biodynamic feedthrough (BDFT)**. The control of many different vehicles is known to be vulnerable to BDFT effects, such as that of helicopters, aircraft, electric wheelchairs and hydraulic excavators.

The fact that BDFT reduces comfort, control performance and safety in a wide variety of vehicles and under many different circumstances has motivated numerous efforts into measuring, modeling and mitigating these effects. Despite the attention that BDFT has received over the last decades, many questions regarding its occurrence remain unanswered. Over the years it has become clear that BDFT is a complex phenomenon in which many different factors play a role. Furthermore, the influence of many of these factors is only poorly understood. It is known that BDFT dynamics depend on vehicle dynamics and control device dynamics, but also on factors such as seating dynamics, disturbance direction, disturbance frequency and the presence of seat belts and arm rests.

The most complex and influential factor in BDFT is the human body. It is through the human body dynamics that the vehicle accelerations are transferred into involuntary limb motions and, consequently, into involuntary control inputs. Human body dynamics vary between persons with different body sizes and weights, but also within one person over time. It is well-known that people adapt their body's neuromuscular dynamics through muscle co-contraction and modulation of reflexive activity in response to, e.g., task instruction, workload and fatigue. This renders BDFT a variable dynamical relationship, not only varying between different persons (between-subject variability), but also within one person over time (within-subject variability).

The research goal of the work presented in this thesis was **to increase the understanding of BDFT to allow for effective and efficient mitigation of the BDFT problem**. This thesis deals with several aspects of biodynamic feedthrough, but the work focused on **the influence of the variable neuromuscular dynamics on BDFT dynamics**. The approach of the research consisted of three parts: first, a method was developed to accurately measure BDFT. Then, several BDFT models were developed that describe the BDFT phenomenon based on various different principles. Finally, using the insights from the previous steps, a novel approach to BDFT mitigation was proposed.

In order to gain a proper understanding of the dependency of BDFT on neuromuscular dynamics, both need to be measured simultaneously. A good measure to describe neuromuscular dynamics has proven to be the neuromuscular admittance. Admittance is a dynamic property of a limb, characterized by the relationship between force input and position output of a limb. This thesis proposes a measurement method that allows for measuring BDFT dynamics and neuromuscular admittance simultaneously. Using this method, insights were gained regarding the relationship between these two dynamics. The thesis describes the method in detail and presents results of experiments in which the method was validated.

In the experiments the admittance was varied using three distinct control tasks: a position task (PT) or 'stiff task', with the instruction to minimize the position deviation of the control device, a force task (FT) or 'compliant task', with the instruction to minimize the force applied to the control device, and a relax task (RT), with the instruction to relax the arm while holding the control device. By following the PT, FT and RT instructions the subject attained respectively a maximally stiff setting of the neuromuscular system (a low admittance), a maximally compliant setting (a high admittance), and a passive setting.

The results of the experimental validation of the proposed method showed that the method was successful in the simultaneous measurement of admittance and BDFT. Based on the observed variations in BDFT dynamics and neuromuscular admittance it was concluded that **there exists a strong dependency of BDFT dynamics on neuromuscular admittance**.

In the literature there is little consensus on how to approach biodynamic feedthrough problems in terms of definitions, nomenclature and mathematical descriptions. This thesis proposes a framework for BDFT analysis which aims to provide a common ground to study, discuss and understand BDFT and its related problems. Using this framework, old and new BDFT research can be (re)interpreted, evaluated and compared. Also, and equally important, the framework itself allows for gaining new insights into the BDFT phenomenon.

Within the framework, a distinction is made between the effects of BDFT on the generation of involuntary control forces and on the generation of involuntary control device deflections (positions). It is proposed to label them BDFT to forces (B2F) and BDFT to positions (B2P) respectively. In addition to B2P, which is the focus of most existing BDFT literature, B2F provides valuable insights. The B2F dynamics can be defined in two different ways, giving rise to the terms BDFT to forces in open-loop (B2FOL) and BDFT to forces in closed-loop (B2FCL). Both forms of B2F dynamics describe different aspects of the BDFT phenomenon.

The framework also includes mathematical relationships, describing how different dynamics relate to each other. The proposed relationships were validated using experimental data. The conclusion following from this was that **the framework proved to be useful in both interpreting previous BDFT studies and in gaining new insights.**

The currently existing BDFT models can be roughly divided into two groups: physical BDFT models and black box BDFT models. Both aim to describe BDFT dynamics, but through different modeling approaches. Physical models are geared towards providing a physical representation of the BDFT phenomenon, using a-priori knowledge and physical principles. Black box models aim to provide an efficient BDFT description at 'end-point level'. In this thesis, two novel BDFT models are proposed. The first is a physical model based on neuromuscular principles, which serves primarily the purpose of increasing the understanding of the relationship between admittance and BDFT. The second model is a mathematical model, which aims to fill the gap between the traditional physical and black box models.

For the first model, a validation demonstrated that **the physical BDFT model provides an accurate physical description of the BDFT dynamics, increasing our fundamental understanding of the BDFT phenomenon.** One of the major contributions of this model is its capability to describe both between-subject and within-subject BDFT variability, something that is often not included in existing BDFT models.

The second model, the mathematical BDFT model, was constructed using asymptote modeling, which offers a structural method to design a model's transfer function. The result is a highly accurate BDFT model of limited complexity, which allows for a reliable parameter estimation and a straightforward implementation. A study of the model's performance led to the conclusion that **the mathematical BDFT model is highly accurate, outperforms several black box models and is easier in use than a physical model.** Furthermore, it was concluded that **asymptote modeling proved to be successful in obtaining an accurate and versatile model structure for the mathematical BDFT model.** The method is likely to be useful in other modeling problems as well.

Next, the issue of BDFT mitigation is addressed. Using the BDFT system model, the available BDFT mitigation techniques were listed and evaluated. A total of seven different solution types, each providing one or more solution approaches, were identified and discussed. Two solution types were deemed most promising. Measures of the first solution type – passive support/restraining systems (e.g., seat belts and armrests) – are already commonly applied. Studies have shown that these are not sufficient to remove BDFT completely. The second promising solution type is model-based BDFT cancellation, where use is made of a BDFT model to obtain a canceling signal. This approach has received some attention in the literature, but only very few experimental implementations have been reported. Using a method called optimal signal cancellation it is shown that **signal cancellation is only a promising mitigation method for BDFT problems if the model can be adapted to both subject and task**. Adaptation to task, or more correctly, to the neuromuscular dynamics of the human operator, is of particular importance.

The effectiveness of using an armrest, an example of a passive support/restraining system, in mitigating BDFT was experimentally investigated. The results show that, generally, the presence of an the armrest reduces the occurrence of BDFT. The results furthermore provide the novel insight that the effect of the armrest varies strongly with frequency and neuromuscular admittance. The main finding of the analysis is that **an armrest is an effective tool in mitigating biodynamic feedthrough**. The installation of an armrest may very well be sufficient to obtain adequate task performance and prevent closed-loop oscillations in many situations that are currently suffering from the occurrence of BDFT. This makes an armrest a viable alternative to more complex mitigation methods.

Finally, a novel approach to BDFT mitigation is proposed: admittance-adaptive model-based signal cancellation. What differentiates this method from other BDFT mitigation approaches available in the literature is that it accounts for adaptations in the neuromuscular dynamics of the human body. The approach was tested, as proof-of-concept, in an experimental setup where subjects inside a motion simulator were asked to fly a simulated vehicle through a virtual tunnel. By evaluating the performance with and without motion disturbance active and with and without cancellation active, the performance of the cancellation approach was evaluated. Results showed that **the admittance-adaptive model-based signal cancellation approach was successful and largely removed the negative effects of BDFT on the control performance and control effort**. From a synthesis of the results presented in this thesis, the following general conclusions can be drawn:

- The current BDFT research environment is fragmented and BDFT problems are often investigated on a case-to-case basis. An increased consensus in the definitions, nomenclature and mathematical descriptions would benefit the understanding of biodynamic feedthrough, improve the communication between researchers and facilitate the comparison between studies.

- Neuromuscular dynamics, and especially the variability thereof, have an important influence that needs to be accounted for when measuring, modeling and mitigating biodynamic feedthrough effects.

- There are many possible ways in which biodynamic feedthrough can be modeled. The preferred way strongly depends on the intended use of the model. In general, biodynamic feedthrough models should be designed such that specialists can incorporate novel insights with the necessary degree of detail, while retaining sufficient practical usability to allow the result to be used by a larger user community.

- The mitigation of biodynamic feedthrough through model-based signal cancellation is a powerful and versatile approach, but only successful if the biodynamic feedthrough model is adapted to both human operator and control task. More research is required to overcome the obstacles that currently prevent the application of model-based signal cancellation in actual vehicles.

Finally, for future research the following recommendations are made:

- Novel identification methods need to be developed to facilitate the measurement of BDFT dynamics in more natural control tasks.

- The further use of the framework for BDFT analysis that was proposed in this thesis should be encouraged in order to improve and extend it.

- Future BDFT modeling efforts should be directed at obtaining models of limited complexity, without compromising fidelity.

- Practical research investigating the occurrence of BDFT in actual vehicles is indispensable to find effective solutions to the BDFT problems occurring around us.

Nomenclature

Acronyms

APC	Aircraft-Pilot-Coupling
ARMA	Autoregressive moving average
ARMAX	Autoregressive moving average with exogenous inputs
AVIS	Active Vibration Isolation System
B2A	Biodynamic feedthrough to accelerations
B2F	Biodynamic feedthrough to forces
B2FCL	Biodynamic feedthrough to forces in closed-loop
B2FOL	Biodynamic feedthrough to forces in open-loop
B2P	Biodynamic feedthrough to positions
BDFT	Biodynamic feedthrough
BMI	Body mass index
CD	Control device
CDFT	Control device feedthrough
CE	Controlled element

CL	Closed-loop
CNS	Central nervous system
COM	Crossover Model
DOF	Degree of freedom
DUT	Delft University of Technology
EMG	Electromyography
FC	Force cancellation
FDFT	Force disturbance feedthrough
FT	Force task
GTO	Golgi tendon organ
HFA	High frequency asymptote
HITS	Highway-in-the-sky
HO	Human operator
HOCD	Human operator - control device interface
IMU	Inertial Measurement Unit
LAT	Lateral
LCD	Liquid-crystal display
LFA	Low frequency asymptote
LMS	Least mean square
LNG	Longitudinal
MC	Motion condition
MSD	Mass-spring-damper
NMS	Neuromuscular system
OL	Open-loop
OSC	Optimal Signal Cancellation

PAO   Pilot-Assisted/Augmented-Oscillation

PATS   Personal Air Transport System

PAV   Personal Aerial Vehicle

PIO   Pilot-Induced-Oscillation

PLF   Platform

PLFHO   Platform - human operator interface

PSD   Power spectral density

PT   Position task

RF   Ratio function

RMS   Root-mean-square

RPC   Rotorcraft-Pilot-Coupling

RT   Relax task

SC   Signal cancellation

SC   Static condition

SD   Standard deviation

SEM   Standard error of the mean

SNR   Signal-to-noise ratio

SRS   SIMONA Research Simulator

TSK   Task

VAF   Variance Accounted For

VRT   Vertical

Greek Symbols

β Relative damping of muscle dynamics   [−]

Δt Simulation time step   [s]

γ	Order of base function	[–]
$\hat{\Gamma}^2_{adm}$	Squared coherence of the admittance estimate	[–]
$\hat{\Gamma}^2_{B2P}$	Squared coherence of the B2P estimate	[–]
μ	Average	
μ_{ϕ_e}	Average heading error	[rad or deg]
ω_f	Frequencies where force disturbance signal has power	
ω_m	Frequencies where motion disturbance signal has power	
ω_n	Natural frequency of base function	[rad]
ϕ	Vehicle roll angle	[rad]
ψ	Vehicle heading angle	[rad]
ψ_{cur}	Current vehicle heading angle	[rad]
ψ_{tar}	Target vehicle heading angle	[rad]
σ	Standard deviation	
θ_{can}	Control device deflection angle after cancellation	[rad]
θ_{cd}	Control device deflection angle (position)	[rad or deg]
θ_{cd}^{mod}	Modeled control device deflection angle	[rad]
ζ	Damping factor of base function	[–]

Latin Symbols

$\hat{H}_{adm}$	Estimate of the neuromuscular admittance dynamics
$\hat{H}_{B2FCL}$	Estimate of the B2FCL dynamics
$\hat{H}_{B2FOL}$	Estimate of the B2FOL dynamics
$\hat{H}_{B2P}$	Estimate of the biodynamic feedthrough to positions dynamics
$\hat{S}_{\theta,\theta}$	Estimate of the auto-spectral density of $\theta_{cd}(t)$
$\hat{S}_{f,f}$	Estimate of the auto-spectral density of $F_{app}(t)$

$\hat{S}_{fdist,\theta}$	Estimate of the cross-spectral density between $F_{dist}(t)$ and $\theta_{cd}(t)$	
$\hat{S}_{fdist,fdist}$	Estimate of the auto-spectral density of $F_{dist}(t)$	
$\hat{S}_{fdist,f}$	Estimate of the cross-spectral density between $F_{dist}(t)$ and $F_{app}(t)$	
$\hat{S}_{mdist,\theta}$	Estimate of the cross-spectral density between $M_{dist}(t)$ and $\theta_{cd}(t)$	
$\hat{S}_{mdist,mdist}$	Estimate of the auto-spectral density of $M_{dist}(t)$	
τ_{ref}	Reflexive time delay	[s]
b_{arm}	Intrinsic arm damping	[Nms/rad]
b_{cd}	Control device damping	[Nms/rad]
b_{grip}	Grip damping	[Nms/rad]
b_{up}	Lumped/effective damping of the upper body	[Nms/rad]
CF	Crest factor	[–]
d	External disturbance on controlled element	
$E_{\dot{\theta}_{res}}$	RMS of the derivative of the residual control device deflection	[rad/s]
f_k	Frequency of HITS describing sinusoid	[–]
f_{act}	Muscle activation cut-off frequency	[Hz]
F_{app}	External force applied on control device	[N]
F_{arm}	Force applied by the human operator (here, through arm)	[N]
F_{arm}^{rem}	Force applied as result of operator remnant	[N]
F_{b2fol}	Force due to B2FOL dynamics	[N]
F_{cdft}	Control device feedthrough force	[N]
F_{dist}	Force disturbance signal	[N]
F_{nms}	Force applied by the neuromuscular system	[N]

f_n	Natural frequency of base function	[Hz]
F_{tot}	Total force applied on control device	[N]
H_B	Base function dynamics	
H_{act}	Muscle activation dynamics	
H_{adm}	Neuromuscular admittance dynamics	
H_{arm}	Arm dynamics	
H_{B2FCL}	Biodynamic feedthrough to forces in closed-loop dynamics	
H_{B2FCL}^{mod}	Model of B2FCL dynamics	
H_{B2FOL}	Biodynamic feedthrough to forces in open-loop dynamics	
H_{B2F}	Biodynamic feedthrough to forces dynamics	
H_{B2P}	Biodynamic feedthrough to positions dynamics	
H_{B2P}^{mod}	Model of B2P dynamics	
H_{CDFT}	Control device feedthrough dynamics	
H_{CD}	Control device dynamics	
H_{cd}	Control device dynamics	
H_{CE}	Controlled element dynamics	
H_{CNS}	Central nervous system dynamics	
H_{FDFT}	Force disturbance feedthrough dynamics	
H_{grip}	Grip dynamics	
H_{GTO}	Golgi-tendon organ dynamics	
H_{HOCD}	Human operator - control device interface dynamics	
H_{HO}	Human operator dynamics	
H_{ms}	Muscle spindle dynamics	
H_{NMS}	Neuromuscular system dynamics	

H_{PLFHO}	Platform - human operator interface dynamics	
H_{PLF}	Platform dynamics	
H_{up}	Upper body dynamics	
I_{arm}	Endpoint inertia of the arm	[Nms^2/rad]
I_{cd}	Control device inertia	[Nms^2/rad]
I_{up}	Lumped/effective inertia of the upper body	[Nms^2/rad]
j	Imaginary unit	[–]
K	Gain of base function	[–]
K_ϕ	Vehicle roll angle gain	[–]
K_s	Tunnel scaling gain	[–]
k_{arm}	Intrinsic arm stiffness	[Nm/rad]
k_{cd}	Control device stiffness	[Nm/rad]
k_f	Golgi-tendon organ feedback gain	[–]
k_{grip}	Grip stiffness	[Nm/rad]
k_p	Muscle spindle stretch feedback gain	[Nm/rad]
k_{up}	Lumped/effective arm stiffness of the upper body	[Nm/rad]
k_v	Muscle spindle stretching rate feedback gain	[Nms/rad]
m_{arm}	Effective mass of the lower arm	[kg]
m_{cd}	Effective mass of the control device	[kg]
M_{dist}	Motion disturbance signal	[m/s^2]
m_{up}	Effective mass of the upper arm	[kg]
N	Number of measurement samples	[–]
n_{cog}	Cognitive (voluntary) control signal	
p_k	Phase shift of HITS describing sinusoid	[deg]
p_{arm}	Scalar converting lower arm inertia to lower arm mass	[rad/m^2]

P_{can}	Cancellation percentage	[%]
p_{up}	Scalar converting upper arm inertia to upper arm mass	$[\text{rad}/\text{m}^2]$
PSD	Power spectral density function	
RF_{adm}	Ratio function of the admittance dynamics	[–]
RF_{B2FCL}	Ratio function of the B2FCL dynamics	[–]
RF_{B2FOL}	Ratio function of the B2FOL dynamics	[–]
RF_{B2P}	Ratio function of the B2P dynamics	[–]
RF_{FDFT}	Ratio function of the FDFT dynamics	[–]
t	Time	[s]
t_d	Delay between motion command and BDFT response	[s]
VAF	Variance accounted for	[%]
x	Tunnel x-coordinate	[m]
y_{cur}	Current state of controlled element	
y_{err}	Difference between goal and current state of controlled element	
y_{goal}	Goal state of controlled element	
z	Tunnel z-coordinate	[m]

Subscripts

θ	Control device deflection related
app	Applied
arm	Arm related
$cdft$	Control device feedthrough related
cd	Control device related
$comb$	Combined

cur	Current
$dist$	Disturbance related
err	Error
$fdist$	Force disturbance related
$goal$	Goal
$mdist$	Motion disturbance related
nms	Neuromuscular system related
tot	Total

Superscripts

$*$	Complex conjugate
$+$	Uncorrected for CDFT
bk	Stiffness and damping related dynamics
cog	Contribution of voluntary/cognitive inputs to the signal
$comp$	Dynamics obtained with compliant control device dynamics
$conv$	Converted dynamics (from compliant to stiff)
$Fdist$	Contribution of the force disturbance to the signal
I	Inertia related dynamics
$Mdist$	Contribution of the motion disturbance to the signal
mod	Model output
m	Mass related dynamics
$nosup$	Without armrest support
rem	Contribution of operator remnant to the signal
res	Contribution of the residual noise to the signal
$stiff$	Dynamics obtained with stiff control device dynamics

^{sup}	With armrest support

Other

(t)	Continuous-time variable
$\bar{}$	Modeled
$\dot{}$	First time derivative
$\hat{}$	Estimated
[deg], $^\circ$	Degrees
[Hz]	Hertz
[kg]	kilogram
[m]	meters
[N]	Newton
[rad]	radians
[s]	seconds
Freq.	frequency

Contents

CHAPTER

Introduction

THIS thesis deals with an ordinary, everyday phenomenon, experienced by regular people on a regular basis: biodynamic feedthrough (BDFT). Perhaps most surprising about this everyday phenomenon is that most people are unaware of its existence and do not know why it is worth studying. BDFT can be experienced in common vehicles like cars, buses, ships and aircraft. It can also be observed during ordinary daily activities such as walking. What these situations have in common is that the human body is moving with respect to its environment.

When the human body is in motion, or more accurately, in the process of changing motion, it is subjected to accelerations. As the human body has mass, these accelerations give rise to external forces working on the body. These forces are the origin of BDFT problems, as they decrease the accuracy with which we can control the movement of our limbs. When the human body is exposed to accelerations, some accelerations *feed through* the human body and cause involuntary motions of different body parts, such as the head, legs, arms and hands. If this human happens to be engaged in a manual control task, the involuntary limb motions may cause involuntary control inputs. This effect is called biodynamic feedthrough.

As humans are commonly subjected to accelerations and external forces, and are often engaged in manual control tasks at the same time, BDFT effects occur regularly in everyday activities. BDFT is the reason why it is hard to drink coffee while walking or why it is difficult to neatly write a postcard during a bumpy bus ride. The shocks, vibrations and other accelerations that our bodies are regularly exposed to cause involuntary limb motions[a] which deteriorate control performance in manual control tasks.

Biodynamic feedthrough is defined in this thesis as follows:

Biodynamic feedthrough

The transfer of accelerations through the human body during the execution of a manual control task, causing involuntary forces being applied to the control device, which may result in involuntary control device deflections.

Having addressed *what* biodynamic feedthrough is, two interrelated *why* questions require answering: why is BDFT a relevant topic for scientific study? And why is the problem of BDFT not well-known?

To answer the latter question first: the most important reason that many people do not know they are regularly exposed to BDFT is that its effects are often little more than mild annoyances. Humans are highly skilled in adapting to the environment in which a manual control task is to be executed. If this environment happens to include shocks and vibrations, humans can and do adapt their control strategy and behavior to minimize negative effects. In some cases that implies simply avoiding performing tasks that require a certain degree of manual control precision in conditions in which the body is in motion. In other cases, biodynamic feedthrough can be reduced by stabilizing control limbs (e.g., resting the elbow on the knee), by changing limb orientation (e.g., bringing the hands

---

[a]It should be noted that acceleration disturbances influence the human body in various other ways as well [Griffin, 1990; McLeod and Griffin, 1989]. Examples of these are the occurrence of visual impairment (blurring of visual perception) and neuromuscular interference (reduction in signal-to-noise ratio between voluntary and involuntary muscle activity). The work presented in this thesis will only deal with biodynamic feedthrough.

closer to the body), or by relaxing the grip on the control device, amongst other mitigation strategies. As a result, biodynamic feed-through effects are common but, admittedly, often not dramatic in regular manual control tasks.

Another reason for the fact that BDFT is relatively unknown, is that the measures we commonly have at our disposal to mitigate BDFT are not always recognized as such. For example, the comfortable seat of a bus or truck driver does not only improve ride comfort, it also reduces the amount of vehicle accelerations the body of the driver is exposed to. This secondary BDFT mitigating function is much less well-known. Another example is the steadycam: a me-chanical camera stabilizing device. A steadycam system improves the smoothness of camera shots by isolating the camera from a large part of the operator's movements. Steadycam systems are widely known and used, but the fact that the purpose of the steadycam system is to reduce the amount of biodynamic feedthrough is not widely known. This makes biodynamic feedthrough, although com-mon, a relatively unknown problem to the public at large.

This only further increases the relevance of the other why-question: why is BDFT – a relatively unknown and mild annoyance – a rele-vant topic for scientific study? We will see there are actually many situations in which BDFT becomes a critical factor, undermining control performance and endangering safety. To illustrate these cases we need to venture outside of the somewhat mundane exam-ples discussed so far.

1.1 Some illustrative examples

Imagine an ambulance speeding towards the hospital after having picked up a patient in a critical condition. The ambulance is racing through an urban environment, rapidly braking, quickly accelerat-ing and taking turns at high speed. Now consider the paramedic in the back of the ambulance who is attending to the patient. The paramedic wants to administer an intravenous injection to stabilize the patient, a task which obviously requires a certain amount of precision. For a skilled paramedic this task would not be particularly hard, at least not under normal circumstances, that is, when everything is stationary. In this case, however, the same paramedic may face great difficulty in performing this high-precision manual control task, as the accelerations his/her body is exposed to cause difficulties in accurately controlling the position of the needle. In fact, a paramedic on board of a driving ambulance may face difficulties in executing many other manual control tasks as well, such as applying bandage or even filling out a form. The culprit here is biodynamic feedthrough.

A similar situation occurs in a flying rescue helicopter, especially one that is experiencing turbulence. The vehicle accelerations of the helicopter cause involuntary limb motions for all those on board, complicating the execution of manual control tasks. This holds for a paramedic, attending to a patient in the cabin of the helicopter, but also for the pilot in the cockpit. In the latter case, the involuntary limb motions may result in involuntary helicopter control inputs. It does not require much imagination to understand that these may not only reduce the pilot's performance but also jeopardize flight safety.

The occurrence of biodynamic feedthrough in various vehicles has been reported in the literature. For helicopters, BDFT is recognized as a problem for a variety of operations [Gabel and Wilson, 1968; Pavel et al., 2011; Walden, 2007]. An example of BDFT in helicopters is vertical bounce, where vertical accelerations cause involuntary control inputs at the collective pitch stick, leading to an adverse coupling between the vertical motion of the helicopter and the pilot's body (e.g., [Gennaretti et al., 2013; Masarati et al., 2014; Mayo, 1989]).

Also aircraft can suffer from BDFT under various conditions, for example when flying through atmospheric turbulence. In a study investigating the impact of structural vibrations on flying qualities of a supersonic transport aircraft several incidents were encountered where the cockpit vibrations fed back into the control stick through involuntary motions of the pilot's upper body and arm [Raney et al., 2001]. Another example is roll-ratcheting: high-frequency roll oscillation that can occur during roll maneuvers in high-performance (fighter) aircraft. The large accelerations induced by a rolling maneuver can trigger or sustain involuntary inputs resulting in an involuntary roll oscillation [Hess, 1998].

Another example of a vehicle that is prone to BDFT is the hydraulic excavator, where driving maneuvers or boom operations can cause strong cabin accelerations, causing involuntary control inputs [Humphreys et al., 2014]. BDFT is also known to occur in electrically powered wheelchairs, where the fore-and-aft accelerations caused by accelerating and braking may induce sagittal (forward-backward) torso and arm motions. This may adversely affect control performance and lead to a phenomenon called 'bucking': an involuntary oscillation alternating between accelerating and braking [Banerjee et al., 1996; Bennet, 1987].

The fact that biodynamic feedthrough is leading to control problems in many different vehicles makes it a relevant topic for study. The observation that BDFT occurs for professional operators in, e.g., excavators, aircraft and helicopters, indicates that the current solution strategies are not always adequate. This raises the need for different, novel techniques that may help us to reduce BDFT and make the control of vehicles safer and easier. The motivation for the work presented in this thesis is to develop and investigate those techniques.

1.2 Factors in biodynamic feedthrough

Fig. 1.1 – adapted from [Venrooij et al., 2011a] – shows an aircraft pilot flying through turbulence. This figure is used here to introduce the main elements that can be distinguished in a BDFT system. The pilot is referred to as the **human operator (HO)**, manually controlling a **controlled element (CE)** (in this case the aircraft) by using one or more **control devices (CDs)** (in this case the control column). The HO applies forces to the CD through his **neuromuscular system (NMS)**. The NMS represents the dynamics of the limb connected to the CD and contains body parts such as bones, muscles, etc. In this thesis this limb will be assumed to be a human arm. In

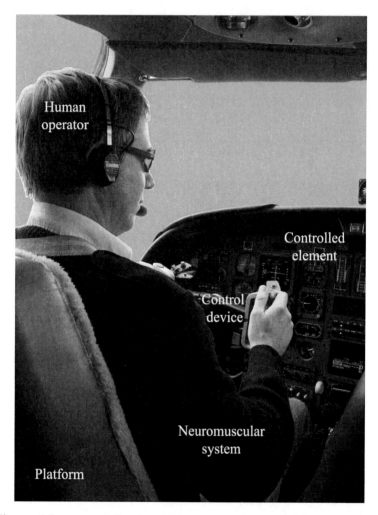

Figure 1.1: An example of a BDFT situation: an aircraft flying through turbulence. The pilot (human operator, HO) is controlling the aircraft (controlled element, CE) by means of the control column (control device, CD). The accelerations of the pilot's seat (platform, PLF) are transferred into the pilot's body (neuromuscular system, NMS). The involuntary motions of the pilot's arms result in involuntary control inputs.

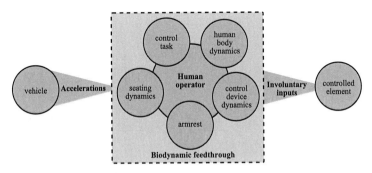

Figure 1.2: A schematic illustration of the biodynamic feedthrough problem. Central to any occurrence of BDFT is the human operator. In addition, some selected factors that influence biodynamic feedthrough are indicated.

addition to the *voluntary* control inputs, the pilot may apply *involuntary* control inputs caused by vehicle accelerations. The physical system from which these accelerations originate is referred to in this thesis as the **platform (PLF)**. In the case shown in Fig. 1.1 the PLF is the airframe of the aircraft. A breakdown similar to the one shown Fig. 1.1 can also be made for other situations where BDFT can occur, such as in an excavator or an electric wheelchair.

In the example illustrated in Fig. 1.1 the PLF and CE belong to the same system, namely the aircraft. Distinguishing between the PLF and the CE is important though, as there are many situations in which they do not belong the same system, such as is the case for a paramedic on board of a helicopter. In that case the PLF is the airframe of the helicopter, but the CE may be something completely different, such as a syringe. Still, the accelerations of the PLF cause involuntary limb motions, complicating the precise control of the CE in very similar ways. The latter situation, in which the HO has no influence on the motions of the PLF, is called an **open-loop (OL)** BDFT system, the former situation is called a **closed-loop (CL)** BDFT system.

Fig. 1.1 provides an intuition regarding which factors play a role in the occurrence of BDFT. For example, the dynamics of the aircraft determine the acceleration the pilot is exposed to, the pilot's seat influences the feedthrough of vehicle accelerations into the pilot's body, the control device dynamics have an effect on how involuntary applied forces result in involuntary control device deflections, etc. Fig. 1.2 shows a highly simplified diagram, illustrating the biodynamic feedthrough problem. Central to any occurrence of BDFT is the human operator. In addition, some selected factors that influence biodynamic feedthrough are indicated (in smaller circles). Next to the factors that were already mentioned – the vehicle dynamics, control device dynamics and seating dynamics – several others factors are indicated. For example, the presence of an armrest, which allows the HO to stabilize the arm. Also, the controlled element (*what* the HO is controlling) and the control task (*how* the HO is controlling) play a role in the occurrence of BDFT. The figure is in no way intended to be exhaustive. There are many factors – even some that are addressed in this thesis – that are not listed in Fig. 1.2.

A factor that is highly influential for BDFT, and particularly relevant for this thesis, is the human body dynamics of the HO, i.e., the dynamics of the neuromuscular system. If this body is big and heavy the neuromuscular dynamics are different than when this body is small and light. If the HO is stressed and 'stiffens up', his body dynamics (and thus the BDFT dynamics) differ from those when the HO is relaxed. Also, how the HO grips the control device, firmly or loosely, plays a role in the occurrence of BDFT.

The human body dynamics can be described by the so-called **neuromuscular admittance**. Admittance is the dynamic relationship between the force acting on the limb (input) and the position of the limb (output). A small admittance implies that a force acting on a limb results in small position deviations, which would occur for stiff limbs. Conversely, a large admittance implies that a force results in large position deviations, which would occur for compliant limbs. It is important to note that humans are capable of varying the neuromuscular admittance of their limbs through muscle activity and reflexive activity. It will be shown in this thesis that these variations have an influence on biodynamic feedthrough dynamics. The relationship between neuromuscular admittance and biodynamic feedthrough has so far not been systematically studied in the literature and is one of the main topics that is investigated in this thesis.

1.3 The complexities of biodynamic feedthrough

Previous research has shown that BDFT is a highly relevant problem in the control of vehicles and machines, but also that it is still poorly understood, both its fundamentals and its practical occurrences. Here we will introduce what can be considered to be the four main complexities of the BDFT phenomenon.

The **first complexity** was already briefly mentioned and is that *biodynamic feedthrough is influenced by a large number of different factors which are often only poorly understood*. Amongst the influencing factors are the acceleration direction, magnitude and frequency content [Lewis and Griffin, 1978a,c; McLeod and Griffin, 1989; Venrooij et al., 2011a]; control device type, position and dynamics [Lewis and Griffin, 1978c; McLeod and Griffin, 1989]; seating dynamics and restraints such as seat belts or armrest [McLeod and Griffin, 1995; Schoenberger and Wilburn, 1973; Torle, 1965; Venrooij et al., 2012]; and the system dynamics which are being controlled. Although progress has been made in the last decades, we are only beginning to understand how these different factors influence the BDFT dynamics.

In the large spectrum of factors that play a role in the occurrence of BDFT there is one that can be considered to be the most influential, complex and poorly understood: the influence of the body dynamics of the human operator. These human body dynamics are of importance as they largely determine whether and how accelerations result in the involuntary limb motions, which in turn are the source of BDFT problems. The human body dynamics vary between persons, for example due to different body sizes and weights, and also within one person over time. Humans adapt their neuromuscular dynamics through muscle co-contraction (the simultaneous contraction of agonist and antagonist muscles) and modulation of reflexive activity in response to, e.g., task instruction, workload and fatigue

[Abbink and Mulder, 2010; Lewis and Griffin, 1978c; Mulder et al., 2011]. This gives rise to the **second complexity** of BDFT: *biodynamic feedthrough is a variable dynamical relationship, varying both between different persons (between-subject variability), as well as within one person over time (within-subject variability)*.

Another factor that complicates studying BDFT is the **third complexity**: *BDFT occurs for a large range of vehicles, under many different circumstances*. Although many of the underlying mechanisms in the occurrence of BDFT across different vehicles are the same, each vehicle has its own peculiarities. Finding commonalities in the BDFT occurrences across vehicles is a challenging task. As a result, most BDFT studies have been devoted to a particular occurrence of BDFT in a specific vehicle, providing results and insights that are vehicle-specific and often even case-specific.

In the past decades, the focus of BDFT research has scattered across the large scope of possible points of interest. This has led to a limited consistency across the various BDFT studies. For example, several different identifiers have been used to describe the BDFT phenomenon, such as vibration breakthrough [Jex and Magdaleno, 1978; Lewis and Griffin, 1976; McLeod and Griffin, 1989], biodynamic coupling [Idan and Merhav, 1990] and operator-induced oscillations [Sirouspour and Salcudean, 2003], amongst other combinations of the adjectives 'vibration(-induced)', 'motion(-induced)', 'biomechanical' or 'biodynamic' with the nouns 'feedthrough', 'feedback', 'interference' or '(cross-)coupling'. In the aeronautical domain, where BDFT has been vigorously studied, various different terms are used for BDFT related effects, such as Pilot-Assisted-Oscillations (PAOs) [Mayo, 1989] and Aircraft- or Rotorcraft-Pilot-Couplings (A/RPCs) [Hamel, 1996; National Research Council, 1997]. For these effects it is known that BDFT can trigger or sustain an adverse coupling. Furthermore, some examples of BDFT effects in particular vehicles have received their own names, such as 'roll ratcheting' (the high-frequency oscillation in high-performance aircraft [Hess, 1998]), 'bucking' (the fore-and-aft oscillation in electrically powered wheelchairs [Bennet, 1987]) and 'human hunting' (coined in a study regarding BDFT effects in excavators [Arai et al., 2000]).

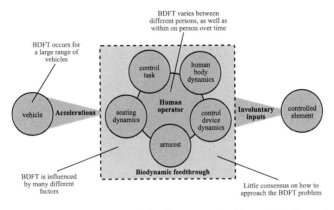

Figure 1.3: A visualization of the four complexities of biodynamic feedthrough.

The sheer abundance of different names used in the literature referring to the same or at least similar phenomena impedes a clear communication between researchers and comparison between studies. This amounts to the **fourth complexity** of biodynamic feedthrough: *there is little consensus on how to approach the BDFT problem in terms of definitions, nomenclature and mathematical descriptions.*
Fig. 1.3 provides a visualization of the four complexities that were just introduced.

1.4   Previous biodynamic feedthrough studies

Biodynamic feedthrough research has been performed for many decades. Instead of providing an extensive discussion of all the individual works that have influenced the research presented in this thesis, this section presents a generalized overview of how the BDFT research field has developed in the past decades, mainly focusing on studies that are not mentioned elsewhere in this thesis. This is followed by five conclusions drawn from the general body of available literature. These conclusions serve to understand the context of the work presented in the current thesis.

Around the 1950s and 1960s the field of biodynamics saw a tremendous increase in research activity, mainly motivated by the advances in aviation and space exploration. A study by Brown and Lechner [1956] stated that the "growing interest in such problems as the optimum flight path to be used [in] man-made satellites [...] has created a need for information concerning the effects of accelerations on man's ability to perform various control functions and on his performance capabilities in general under a variety of acceleration conditions". Furthermore, it was remarked that "it is essential to consider the mechanical effects of accelerations on movements of various parts of the body" [Brown and Lechner, 1956]. That the importance of biodynamics in manned space flights did not go unnoticed is signified by the official nomenclature of the Apollo space program, which defined "biodynamics" as "[the] study of forces acting upon bodies in motion or in the process of changing motion, as they affect living beings" [Anon., 1963]. The immense bibliographies regarding biodynamic research that appeared around that time, such as [Snyder et al., 1963] and [Jones, 1971], each with hundreds of entries, are a further testimony to this increased interest. A portion of these studies reported or investigated the biodynamic feedthrough phenomenon (although the term 'biodynamic feedthrough' itself did not appear until much later). An example of one of these earlier studies is the study by Larue [1965], which investigated the degradation of accuracy that occurred in a positioning task when performed in a vibration environment.

In the 1970s, some aspects of human control behavior were already well understood (such as the effect of the forcing function), but the understanding of the influence of many environmental variables (such as vibrations) was still quite limited. Jex [1972] provided an overview of some biodynamic man-machine control problems and their status, illustrating amongst other things that in the early 1970s only limited knowledge was available on the effect of vibration on tracking performance. Much research effort has since been devoted to identifying the different sources of the vibration-induced errors and their relative importance. Many studies used sinusoidal vibrations to uncover the mechanisms of BDFT and related phenomena on tracking performance. The extensive research reported in [Allen et al., 1973] was largely geared towards partitioning human control behavior into three parts: voluntary/involuntary visual-motor response, the contribution of vibration feedthrough (which is called biodynamic feedthrough in the current thesis), and the remaining portion, or remnant, which is uncorrelated with either command or vibration inputs. The remnant can be further divided into several contributions, such as perceptual remnant (caused by visual blurring) and motor remnant (caused by interference with neuro-muscular actuation processes). A similar partitioning was used in several other studies, such as [Lewis and Griffin, 1976], where it was suggested that interference with the kinaesthetic feedback mechanisms (i.e., disruption of the information provided by receptors in the limbs) may be the principal means by which vibration degrades tracking performance. This hypothesis was later partially contradicted by the results obtained in [Lewis and Griffin, 1979].

In a review of the effects of vibration on visual acuity and continuous manual control [Lewis and Griffin, 1978b,c], summarizing the available knowledge at the end of the 1970s, it was stated that "despite the range of interest and information that has been covered [...], most of the general conclusions which can be drawn from the results of the research are not very far reaching and none are without disagreement" ([Lewis and Griffin, 1978c], p. 415).

One of the most controversial issues at the time was the effect of duration of vibration exposure. Some studies showed a dependency between the duration of exposure to vibration and tracking performance, while others did not (see the summary table in [McLeod and Griffin, 1989]). Some evidence suggested that a moderate exposure up to several hours increased arousal, partially mitigating the effect of fatigue and generally improving tracking performance [Lewis and Griffin, 1978c, 1979]. When the question of the effects of vibration exposure duration on control performance was revisited more than a decade later in [McLeod and Griffin, 1993] the controversy was not yet solved, as it was found that the control performance itself was influenced by the duration of the task, but this effect was not altered by the exposure to vibration.

Another source of disagreement between studies was (and in many respects still is) the individual variability in the response to motion disturbances. Griffin and Whitham [1978] reported the vibration response of 120 seated subjects and concluded, amongst other things, that there exists a correlation between physical characteristics and their biodynamic response. It was claimed that this variability may be an important contributing source to the inconsistencies that had appeared in the literature [Griffin and Whitham, 1978].

The considerable progress that was made at the end of the 1980s in the understanding of the effects of motion disturbances on the human body is illustrated in the comprehensive review of BDFT related literature provided in [McLeod and Griffin, 1989]. The review is organized on the basis of topics and variables. This facilitates the comparison of single variables across studies. The paper also provides a behavioral model summarizing the mechanisms by which vibration can interfere with manual control performance. Four major mechanisms are distinguished: vibration breakthrough, visual impairment, neuromuscular interference, and central effects. The current thesis is mainly concerned with the first of these mechanisms.

The term "vibration breakthrough", as used in [McLeod and Griffin, 1989] and other works, can be considered to be largely identical to what is labeled "biodynamic feedthrough" in the current thesis, with one subtle but important difference: vibration breakthrough typically includes the response of the controlled element, while biodynamic feedthrough does not. As defined in the beginning of this chapter, BDFT describes the transfer of accelerations to involuntary control inputs such as control forces and/or control device deflections. This implies that BDFT excludes the controlled element dynamics, as can be observed in Fig. 1.2. The way vibration breakthrough is typically defined is that it describes the transfer of accelerations to controlled element output, which includes the controlled element dynamics. This distinction is of importance to appreciate the results obtained in this thesis and in order to compare the results with those of previous studies (see Chapter 9).

One of the earliest usage of the term "biodynamic feedthrough" was in [Jewell and Citurs, 1984], where it also happened to be defined in the form similar to what we will use in this thesis: "how the aircraft's lateral accelerations [...] affect the pilot's controls", referring to the involuntary control device deflections caused by motion disturbances.

Biodynamic models can be designed for several different purposes [Griffin, 1981], amongst which the prediction and understanding of the effect of vibration on the human body. The biodynamic models that are available in current literature were developed for a multitude of purposes and not all of them deal with biodynamic feedthrough. Biodynamic models can be categorized using three model types [Griffin, 2001]: mechanistic models (also known as physical models [Sövényi, 2005]), quantitative models (also known as black-box models [Sövényi, 2005]) and effects models. The current thesis will only deal with the first two model types.

In the early 1970s, the first physical (mechanistic) BDFT models appeared, e.g., [Allen et al., 1973; Jex and Magdaleno, 1978, 1979]. These models were primarily constructed using a-priori knowledge and physical principles. The models were typically validated using experimental data of the biodynamic response of body parts to vibration disturbances of varying magnitude and frequency (e.g., [Allen et al., 1973; Donati and Bonthoux, 1983; Jex and Magdaleno, 1978]). Results of these studies led to the conclusion that, under certain circumstances, the human body dynamics can be approximated by a linear mechanical system [Donati and Bonthoux, 1983; Lewis and Griffin, 1979].

Examples of black box (quantitative) BDFT models can be found in [Mayo, 1989; Sövényi, 2005; Velger et al., 1984]. As the name already suggests, these models consider the BDFT dynamics as a black box and describe the relationship between input and output without considering the physical elements in between. A large number of different physical and black box BDFT models were developed.

The main advantage provided by physical BDFT models over black box BDFT models is the additional insight gained in the physical processes underlying the BDFT phenomena. This insight often comes at a price: as such models aim to describe the complexities of reality, they are usually more elaborate than their black box counterparts, which merely seek an efficient description of the dynamics at 'end-point level'. The parameter estimation of an elaborate physical model is a challenging task, which may be done faster and more reliably for a black box model. A drawback of the black box models is that their parameters lack a physical interpretation and the models are often more limited to specific applications [Griffin, 2001; Venrooij et al., 2013b].

Over the years, many different types of mitigation techniques were proposed and tested, ranging from simple armrests (passive supports for the forearm) [Torle, 1965] to active vibration isolation systems (active systems that counteract the vibrations of the pilot's chair) [Dimasi et al., 1972; Schubert et al., 1970]. Other studies proposed and tested adaptive filtering techniques [Velger et al., 1984, 1988]. More recently, an approach called model-based BDFT cancellation has been proposed, which relies on a BDFT model to predict and correct for involuntary control inputs (e.g., [Sirouspour and Salcudean, 2003; Sövényi, 2005]). Yet another alternative solution for the occurrence of BDFT, specifically in backhoes (a type of excavator), has been proposed in [Humphreys et al., 2011], which uses the backhoe arm to reduce the cabin vibrations the operator is exposed to. These and other mitigation approaches will be discussed in more detail in Part III of this thesis.

The recent efforts in the BDFT research domain seem to be focused on vehicle specific BDFT occurrences. For example, recent studies have been conducted in the context of the GARTEUR HC-AG16 project (e.g., [Dieterich et al., 2008]) and the ARISTOTEL[b] project (e.g., [Masarati et al., 2013; Pavel et al., 2012; Quaranta et al., 2013]). These projects investigated Rotorcraft-Pilot Couplings (RPCs). It is known that biodynamic feedthrough can both cause and sustain such events. The recent interest in RPCs is driven by the fact that the implementation of more advanced flight control systems in modern helicopters appears to have caused more RPC events than before [Pavel et al., 2011, 2012]. This stresses the need to obtain a more fundamental understanding of how BDFT may interfere with control performance in order to alleviate its effects on RPCs.

Another example of a current European project that considers BDFT

---

[b]http://aristotel-project.eu/

as one of its research topics is the myCopter^c project, which investigates the enabling technologies required for a Personal Air Transport System (PATS) based on Personal Aerial Vehicles (PAVs). One of the work packages of the myCopter project concerns the occurrence of BDFT in PAVs. The work presented in this thesis was partially performed within the context of both the ARISTOTEL and the myCopter project.

From the available literature on BDFT a number of conclusions can be drawn. The following were the most influential for the work presented in this thesis:

- Previous and current research efforts show an ongoing interest in the problem of biodynamic feedthrough and related topics.

- The fragmentation in BDFT literature obstructs comparing results obtained in existing studies and building on those studies to further our understanding of BDFT phenomena.

- There is not one accepted model that is used across disciplines to model BDFT phenomena.

- Apart from the use of armrests – of which the exact influence on BDFT is not yet quantified – there is no practical BDFT mitigation method available.

- The influence of human body dynamics has only received limited attention in the literature. These dynamics are highly variable and are likely to play a large role in BDFT. The details of this interaction are, however, poorly understood and have not been systematically studied.

1.5   Motivation, goal and approach

What seems to be lacking in current BDFT literature is a systematic study into the variability introduced by the human body dynamics. As human body dynamics are of importance for *any* occurrence of

---

^c http://www.mycopter.eu/

BDFT, increasing the knowledge on its role benefits the understanding of BDFT across a diverse range of situations.

Next to the scientific motivation to increase our knowledge about the fundamentals of the BDFT phenomenon, also a more practical motivation played a role: the motivation to solve this problem. A solution to BDFT would not only make the operation of several vehicles easier and safer, it may also pave the way for new developments in the field of human-machine systems design. For example, a solution to BDFT would allow light-weight intuitive control devices to be used to its full potential, despite their possible susceptibility to BDFT effects. It may help to make electric wheelchairs become more agile, improving the quality of life of their owners, without increasing the occurrence of bucking effects. Also, it may reduce the training time needed for novice PAV pilots, as the involuntary inputs due to the accelerations are automatically canceled. Hence, an increased knowledge of BDFT does not only solve some of our current problems, it may also provide access to yet unexplored opportunities.

The goal of the research work that is presented in this thesis was:

Research goal

Increase the understanding of BDFT to allow for effective and efficient mitigation of the BDFT problem.

The research conducted in the context of this thesis focused mainly on the influence of the human body dynamics on BDFT, as this is the area where an increase in knowledge seems most needed. However, as we will see in the following chapters, the insights gained with respect to the influence of the human body dynamics have led to additional insights into other influencing factors as well. Effective and efficient mitigation should be understood as mitigation that largely removes the involuntary BDFT inputs (effective) without involving high costs (efficient). Costs refer not only to financial costs, but also amongst other things to the complexity of the approach, the time its implementation requires and the sacrifices it involves in other aspects of vehicle control.

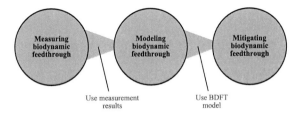

Figure 1.4: The approach taken in this research consisted of three phases: measuring BDFT, modeling BDFT and mitigating BDFT. The measurement results obtained in the first phase were used to develop biodynamic feedthrough models in the second phase. One of those models was used to mitigate biodynamic feedthrough in the third phase.

The approach taken to reach this goal consisted of three general parts, dividing the research performed in this thesis in three distinct phases: measuring biodynamic feedthrough, modeling biodynamic feedthrough and mitigating biodynamic feedthrough. Fig. 1.4 provides a schematic overview of the research approach, which is detailed below.

1.5.1 Measuring biodynamic feedthrough

In order to study the influence of human body dynamics on BDFT dynamics a method needs to be developed to measure both dynamics simultaneously. The human body dynamics can be described by the neuromuscular admittance, which can be structurally varied by using different manual control task instructions. By measuring both neuromuscular admittance and BDFT simultaneously, it can be investigated how changes in the neuromuscular admittance influence the BDFT dynamics. The development and validation of such a method was the central theme in the first phase of the research.

1.5.2   Modeling biodynamic feedthrough

The results obtained from BDFT measurements were used to construct and validate novel BDFT models. These models allow us to gain a deeper understanding of the BDFT problem and provide us with tools to mitigate its effects. Two novel models were developed in the second phase of the research.

Modeling biodynamic feedthrough requires defining which dynamics the model is describing. This calls for some agreement on 'what is what'. The definitions and notation that were developed and used in the first and second phase of the research were combined into a framework for BDFT analysis. This framework defines how different elements that influence BDFT can be called and how their interaction can be mathematically described. Although the development of a framework was not part of the original approach of measuring, modeling and mitigating BDFT, it is a valuable by-product of the research efforts.

1.5.3   Mitigating biodynamic feedthrough

The third phase of the research focused on the question which ways of mitigating BDFT we have at our disposal. And, also, which of these ways are most promising. The range of possible approaches stretches from simple and straightforward solutions to complicated and delicate ones, each having its own benefits and disadvantages. This thesis will investigate two options in more detail, a simple approach – an armrest – and a more complicated one – admittance-adaptive model-based signal cancellation. In this latter approach one of the BDFT models that was developed in the previous research phase was used.

1.6   Scope of thesis

Not every aspect of biodynamic feedthrough could be dealt with within the scope of this thesis. The research scope was limited by the following factors:

- All measurements have been performed in open-loop. This implies that in the experiments there existed no coupling between the control inputs of the human operator and the motions of the motion simulator.

- The disturbances were always applied in one direction, along a single axis only. Hence, possible cross-coupling effects of multi-directional disturbances were not investigated.

- The study mainly focused on BDFT effects in the lateral axis.

- The cognitive/voluntary aspects of human control have not been studied in detail. The study focused on the involuntary BDFT induced control inputs.

The motivation for these choices and their implications for the applicability of the results will be addressed in the following chapters.

1.7   Guidelines for the reader

The chapters of this thesis are largely based on published work. The research that led up to this thesis was published in 11 conference papers and 9 journal papers. Of these, 4 conference papers and 7 journal papers served as a source for the work presented in this thesis. The contents of some papers were merged into single chapters. In several chapters new unpublished material was added. The contents of the chapters have been adapted to increase consistency and improve readability of the thesis as a whole. For example, the nomenclature and notation was unified. Also, several references to papers contained in this thesis were replaced with references to thesis chapters. Many sections were rephrased or rewritten to improve readability.
At the same time, however, it was attempted to maintain the general outline of the original papers. The goal was to retain sufficient information in each chapter to ensure its independent readability. This choice does imply, however, that some material will be repeated throughout the thesis. In most cases this was done deliberately. For example, a (brief) discussion of the BDFT system model, a central

model in this thesis, appears in every chapter. In each chapter, the focus will be laid on a different part of the BDFT system model and the model is therefore used as a convenient point of departure, serving to indicate a chapter's focus within the larger context. Also, experiment descriptions are repeated throughout chapters to clarify how the data of each chapter were obtained. Further repetitions were kept at the minimum required to allow for understanding the methods and results of a chapter, without having to refer to other chapters.

The discussion of biodynamic feedthrough involves a vast number of different types of systems, signals and dynamics, each having its own notation and/or acronym. A detailed discussion of these is provided in Chapter 3 and cannot be fully repeated in each chapter. Instead, the nomenclature that precedes this chapter may help to keep track of the acronyms and symbols used throughout the thesis. Another convenient point of reference is the discussion of the most important concepts that is provided in Appendix A. One can review the fundamentals of biodynamic feedthrough there at any time.

1.8   Outline of the thesis

The thesis is divided in three parts, reflecting the three phases in the research approach. In **Part I** we address a method to measure biodynamic feedthrough. Furthermore, we look into ways of analyzing the results. This is followed by **Part II**, where we look into two different approaches of biodynamic feedthrough modeling, resulting in a physical BDFT model and a mathematical BDFT model. Then, in **Part III** we discuss several aspects of biodynamic feedthrough mitigation. We go through the available mitigation techniques, study the effectiveness of an armrest and we develop and validate a novel BDFT mitigation technique. Finally, we discuss the results and draw conclusions. Fig. 1.5 illustrates the outline of the thesis.

The work presented in this thesis has made the following contributions to existing BDFT literature.

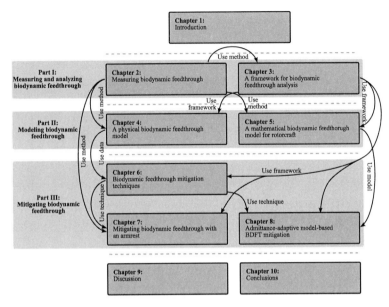

Figure 1.5: A visualization of the outline of this thesis.

1.8.1 A method to measure biodynamic feedthrough dynamics and admittance

Chapter 2 describes the development and experimental validation of a method to measure both BDFT and human body dynamics simultaneously. By extending the methods developed at Delft University of Technology to measure neuromuscular admittance (a reliable measure for human body dynamics), a simultaneous measurement of the BDFT dynamics was realized. This method allows for obtaining measurements of the BDFT dynamics for a range of different settings of the neuromuscular system.

The proposed measurement method was used in several experiments to gain additional insights regarding, e.g., the influence of control device dynamics (Chapter 3), the effect of disturbance direction (Chapter 5), the presence of an armrest (Chapter 7) and the effectiveness of model-based BDFT mitigation (Chapter 8).

1.8.2 A framework to analyze biodynamic feedthrough

Chapter 3 presents a framework for BDFT analysis. This comprehensive set of definitions, nomenclature and mathematical notations provides a common ground to study, discuss and understand BDFT and its related problems. The framework is validated using experimental data. Furthermore, it is illustrated how the framework can be used to gain novel insights into BDFT phenomena and how the approaches and results put forward in previous studies can be reinterpreted. The fact that such a structured framework is not available in existing literature makes this a valuable contribution.

1.8.3 A physical BDFT model

In **Chapter 4** a novel physical BDFT model is developed, which models the physical occurrence of BDFT. The model serves primarily the purpose of increasing the understanding of the relationship between neuromuscular admittance and biodynamic feedthrough. The physical BDFT model provides an unprecedented level of detail regarding the influence of neuromuscular admittance on BDFT dynamics.

1.8.4 A mathematical BDFT model

Chapter 5 describes another BDFT model, an alternative to the physical BDFT model. This model is referred to as the mathematical BDFT model and aims to close the gap between the existing (complex) physical and (limited) black box models. The model was obtained using a new modeling approach, developed within the scope of this thesis, called asymptote modeling. This approach allows for systematically obtaining a model's transfer function structure. Asymptote modeling can also be applied to other modeling problems. The performance of the resulting mathematical BDFT model was validated in both the frequency and time domain. Furthermore, it was compared with several recent BDFT models. The model shows to be highly accurate.

1.8.5   New insights regarding BDFT mitigation

Using the framework for BDFT analysis, all possible BDFT mitigation approaches are identified in **Chapter 6**. From these, two methods are selected for further investigation. First, in **Chapter 7**, the mitigation effectiveness of a widely-used hardware component is studied: an armrest. So far, the exact effect of an armrest on BDFT dynamics had not been experimentally quantified. The results show that an armrest is an effective tool in mitigating biodynamic feedthrough. The second mitigation approach type that was selected for further investigation was model-based BDFT cancellation (see next section).

1.8.6   A new approach to BDFT mitigation

Several insights gained over the course of this research are combined in **Chapter 8**, where a novel approach in BDFT mitigation is proposed: admittance-adaptive model-based signal cancellation. What differentiates this approach from other approaches is that it accounts for adaptations in the neuromuscular dynamics of the human body. The approach was tested, as proof-of-concept, in an experiment where subjects inside a motion simulator were asked to fly a simulated vehicle through a virtual tunnel. By evaluating the performance with and without motion disturbance active and with and without cancellation active, the performance of the cancellation approach is assessed. Results show that the cancellation approach was successful and largely removed the negative effects of BDFT on the control effort and control performance.

Part I

Measuring and analyzing biodynamic feedthrough

CHAPTER 2

Measuring biodynamic feedthrough

In this chapter a method is presented to simultaneously measure BDFT and neuromuscular admittance in a motion-based simulator. The method was validated in an experiment. By applying a force disturbance signal to the control device the admittance was measured, by simultaneously applying a motion disturbance signal to the motion base of the simulator the BDFT dynamics were measured. The results show a dependency of BDFT on neuromuscular admittance, i.e., a change in neuromuscular admittance results in a change in BDFT dynamics. Understanding this dependency is essential in understanding BDFT as a whole.

The contents of this chapter are based on:

Paper title   A Method to Measure the Relationship Between Biodynamic Feedthrough and Neuromuscular Admit-
tance

Authors   Joost Venrooij, David A. Abbink, Mark Mulder, Marinus M. van Paassen, and Max Mulder

Published in   IEEE Transactions on System, Man and Cybernetics
- Part B: Cybernetics, Vol. 41, No. 4, Aug. 2011

2.1 Introduction

It is known that humans can adapt the dynamics of their limbs by adjusting their neuromuscular settings. It is likely that these adaptations have an influence on biodynamic feedthrough (BDFT) dynamics as well, but this influence is currently not well understood and was thus far not systematically studied. This chapter presents a method to simultaneously measure BDFT dynamics and limb dynamics in a motion-based simulator, which allows us to investigate this dependency.

Limb dynamics can be described by neuromuscular admittance, the causal dynamic relation between a force input and a position output of a limb. A large admittance means that a force acting on a limb results in large position deviations, which would occur for compliant limbs; a small admittance means a force results in small position deviations, which would occur for stiff limbs. Humans are capable of varying the neuromuscular admittance of their limbs through, e.g., muscle co-contraction (the simultaneous contraction of agonist and antagonist muscles) and reflexive activity. These variations are likely to have an influence on the biodynamic feedthrough, which makes BDFT a variable dynamical relationship, varying both between different persons (between-subject variability), as well as within one person over time (within-subject variability).

In an experiment, neuromuscular admittance was measured by applying a force disturbance signal to the control device; BDFT was measured by applying a motion disturbance signal to the motion base of the motion simulator. To allow for distinguishing between the operator's response to each disturbance signal, the perturbation signals were separated in the frequency domain. To investigate the impact of neuromuscular adaptation on BDFT dynamics, subjects were asked to perform three different control tasks, each requiring a different setting of the neuromuscular system.

The results of the experiment did not only show that the experimental method was successful in measuring both neuromuscular admittance and BDFT simultaneously, but also that a strong dependency indeed exists of BDFT dynamics on neuromuscular admittance. The measurement method proposed in this chapter allows us to increase our understanding of this poorly understood interaction.

The structure of this chapter is as follows: first the approach and scope of the experimental method is introduced in Section 2.2. This is followed by a discussion of the disturbance signal design in Section 2.3. The proposed method was tested in an experiment, described in Section 2.4. The results of this experiment are presented and discussed in Section 2.5, followed by conclusions in Section 2.6.

2.2   Biodynamic feedthrough system model

This section serves to introduce the biodynamic feedthrough system model, a conceptual model that contains the elements that play a role in BDFT. As the model is more detailed than is required to discuss how BDFT dynamics and admittance can be measured, the model will be first presented here in a somewhat simplified form. The BDFT system model will be revisited and discussed in full detail in Chapter 3.

2.2.1   An introduction to the biodynamic feedthrough system model

Recall the example of a BDFT situation provided in Fig. 1.1: a pilot (or human operator, HO) flying an aircraft (controlled element, CE) through turbulence. The HO is controlling the CE by means of the control column (control device, CD). The accelerations of the pilot's seat (platform, PLF) are transferred into the pilot's body (neuromuscular system, NMS). That situation is schematically represented as a block diagram in Fig. 2.1. The figure is not only applicable to the situation where an aircraft pilot is flying through turbulence, but many other situations where BDFT may occur as well.

Each block in Fig. 2.1 contains dynamics that can be described through a transfer function (indicated with H). In many practical cases the dynamics of the CE, CD and PLF are not (significantly)

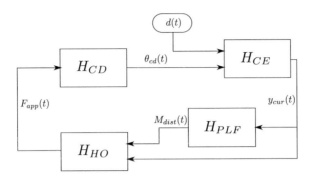

Figure 2.1: A schematic representation of a typical BDFT situation: a human operator (HO) is controlling a controlled element (CE) by means of a control device (CD). The HO is located on a platform (PLF), typically a vehicle or a motion simulator, of which the accelerations can cause BDFT.

varying over time. This leaves the HO as the most important varying element in the BDFT system, i.e., the H_{HO} dynamics can vary over time and between subjects.

The signals between the blocks are labeled: the signal $M_{dist}(t)$ represents the motion disturbance (or acceleration disturbance) signal which perturbs the body of the HO; the HO applies a force $F_{app}(t)$ to the CD, which in response deflects to angle $\theta_{cd}(t)$; the disturbance signal that works on the CE (e.g., turbulence) is indicated as $d(t)$; the current state of the CE is indicated as $y_{cur}(t)$. Note that the HO block has two inputs, the state of the CE $y_{cur}(t)$ and the motion disturbance $M_{dist}(t)$ coming from the PLF. The HO uses the former to formulate voluntary control inputs, the latter is the source of additional involuntary motion-induced control inputs (BDFT).

The challenge that is addressed is the current chapter is how to measure BDFT and neuromuscular admittance simultaneously. An established way of measuring neuromuscular admittance is by using a force disturbance that we will call $F_{dist}(t)$ on the control device. This force disturbance perturbs the limb that is in contact with the control device, allowing us to measure the limb's dynamics.

A complete representation of the BDFT system model is shown in Fig. 2.2. The parts that are not relevant for the current chapter are

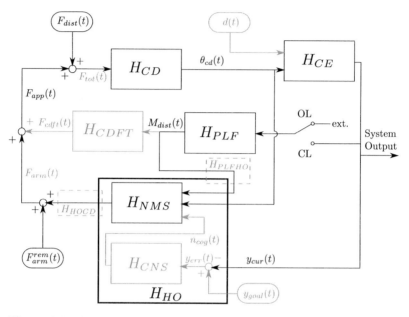

Figure 2.2: The complete biodynamic feedthrough system model. For details refer to the text. Signals and systems that are not relevant for the current chapter are shown in gray.

shown in gray. These will be addressed in detail in Chapter 3. The items shown in black include the blocks and signals that were just introduced, including the NMS as part of the HO and the force disturbance $F_{dist}(t)$. Note that one of the inputs to the NMS, shown in gray, is the signal $n_{cog}(t)$ coming from the central nervous system (CNS). This signal represents the cognitive voluntary control of the NMS.

Two types of BDFT systems can be distinguished: closed-loop (CL) BDFT systems and open-loop (OL) BDFT systems. In terms of the BDFT system model, the difference between the two is whether there exists a connection between the CE and the PLF. In closed-loop BDFT systems, the HO can influence the motion of the PLF through a connection between the CE and the PLF. In the open-loop BDFT systems, the PLF is controlled by external input signals.

In Fig. 2.2 this connection can be opened or closed by a switch.
The reason for including the full BDFT system model in this chapter, although not all aspects are discussed here, is for consistency and future reference. The BDFT system model will appear in each of the following chapters, indicating what the focus of that chapter is. In this case it is developing a method to measure BDFT and neuromuscular admittance simultaneously. This can be done by applying two disturbances: $M_{dist}(t)$ and $F_{dist}(t)$ and measuring two control signals: $F_{app}(t)$ and $\theta_{cd}(t)$.

2.2.2 The occurrence of biodynamic feedthrough

Biodynamic feedthrough occurs when motion disturbances $M_{dist}(t)$ induce unintentional motions in the limb that is in contact with the CD, thereby leading to unintentional control inputs. If a force disturbance $F_{dist}(t)$ is applied in addition to measured admittance, the control input signal $\theta_{cd}(t)$ consists of the following contributions:

$$\theta_{cd}(t) = \theta_{cd}^{Fdist}(t) + \theta_{cd}^{Mdist}(t) + \theta_{cd}^{cog} + \theta_{cd}^{rem}(t) \qquad \text{2.1}$$

where superscript ^{Fdist} denotes the contribution of the force disturbance and ^{Mdist} the contribution of the motion disturbance (the biodynamic feedthrough). The superscript ^{cog} denotes the contribution to the control device deflections that are cognitively applied by the HO. Think for example of the cognitive steering actions during a visual pursuit task. The remaining part of the control input signal, the part that is not related to a disturbance signal or cognition, is the remnant and denoted by the superscript ^{rem}. Remnant can be defined as the operator's control output power that is not linearly correlated with the system input [McRuer and Jex, 1967] (in our case those inputs are the disturbance signals). The applied force $F_{arm}(t)$ can be broken up in the same contributions:

$$F_{arm}(t) = F_{arm}^{Fdist}(t) + F_{arm}^{Mdist}(t) + F_{arm}^{cog} + F_{arm}^{rem}(t) \qquad \text{2.2}$$

In this thesis the cognitive input and remnant, i.e., the contributions that are not related to the two disturbance signals will often be combined in a contribution which will be labeled the residual, indicated with the superscript *res*:

$$F_{arm}^{res}(t) = F_{arm}^{cog} + F_{arm}^{rem}(t)$$

2.3

From Eq. 2.1 and Eq. 2.2 it follows that BDFT effects are present in both the control device deflections and the applied force. To distinguish between these two related effects it is proposed to refer to them as *BDFT to positions* (B2P) and *BDFT to forces* (B2F) respectively (Chapter 3 provides more details on this distinction). The current chapter will only deal with B2P. As the concepts of B2P and B2F are not formally introduced yet, the terms 'BDFT dynamics' and 'B2P dynamics' can be considered synonymous for the length of this chapter.

2.2.3   Scope of research

Biodynamic feedthrough problems can be highly complex. Because the aim of this chapter is to demonstrate a measurement method for general BDFT systems, the complexity of the system considered here has been limited in the following ways:

- The current study focuses on the neuromuscular aspect of BDFT, thus on the role of the NMS of the human operator only; cognitive control actions are not considered. As will be discussed below, the cognitive contribution will be limited by making use of specific control tasks.

- In the analysis, the neuromuscular dynamics and BDFT dynamics are considered to be linear and time-invariant. Evidently, this assumption is not always valid but can be justified for our experiments by carefully crafting the experimental conditions. The validity of this assumption in our experimental setup was checked using the squared coherence of the obtained data.

- In this study only open-loop BDFT systems are investigated. That implies that the human operator has no influence on the PLF motions.

- The investigation deals with the occurrence of BDFT in general and not for any vehicle in particular. Therefore, the dynamics of the PLF or the CE are secondary to our objectives.

- In this study only lateral accelerations are investigated for control tasks using a side-stick.

Using the above considerations, the BDFT system model displayed in Fig. 2.2 can be reduced to a form that is relevant for the current study. As only open-loop systems are considered, the OL/CL switch in Fig. 2.2 is connected to external inputs ('ext.'), over which the HO has no control. No cognitive inputs are considered, so the contents of the CNS block are not investigated in this study. Finally, as the dynamics of the controlled element and platform are secondary to our current goals, also the CE block and the PLF block lie outside the scope of this investigation. What remains from the BDFT system model that is of importance for the current study are the disturbance signals $F_{dist}(t)$ and $M_{dist}(t)$, the NMS block and the CD block.

By investigating Fig. 2.2, the role that is played by the neuromuscular admittance in the occurrence of BDFT becomes apparent. By following the path from motion disturbances $M_{dist}(t)$ to (unintentional) control device deflections $\theta_{cd}(t)$, it can be observed that the dynamics that play a role in the occurrence of BDFT are the NMS dynamics and the CD dynamics.

2.3 Disturbance signal design

2.3.1 Frequency separation of the disturbance signals

Simultaneously applying two disturbance signals $F_{dist}(t)$ and $M_{dist}(t)$ requires a method to distinguish the contribution from each disturbance signal in the measured response. The method used here, and throughout the rest of this thesis, is based on separating the disturbance signals in the frequency domain. By offering each disturbance at a separated set of discrete frequencies, the response to each disturbance signal can be separated in the analysis. This method was successfully employed in several studies to measure

neuromuscular admittance [Abbink, 2006; Damveld et al., 2009].
It is important to note that this method requires the dynamics to be linear time-invariant (LTI). By carefully designing the experimental conditions and task instructions, the highly non-linear and time-variant BDFT dynamics and admittance can be approximated by using these LTI methods.

2.3.2   Reduced power method

It is known that admittance is dependent on the bandwidth of the perturbation signal. It was shown that a low bandwidth perturbation results in a significantly smaller admittance than for perturbations with a larger bandwidth [van der Helm et al., 2002]. It was hypothesized this could be attributed to the suppression of reflexive activity when the limb was excited by higher frequencies. In other words, a different low frequency behavior is measured when higher frequencies are excited, because it causes humans to attenuate their reflexive activity.

To allow estimation of full-bandwidth dynamics, without influencing the low frequency behavior, the reduced power method, elaborated in [Mugge et al., 2007], was used to construct the disturbance signals. This method relies on reducing the power of the disturbance at higher frequencies, such that these frequencies do not influence the control behavior, but still enable estimation of the dynamics at high frequencies.

The reduced power method was used for the design of both the force disturbance signal $F_{dist}(t)$ and the motion disturbance signal $M_{dist}(t)$. It should be noted, however, that the success of the reduced power method has only been demonstrated for admittance measurement. The effect of perturbation bandwidth on BDFT measurements is largely unknown. Additional research is required to show whether and how BDFT measurements are affected by factors such as perturbation amplitude and frequency content.

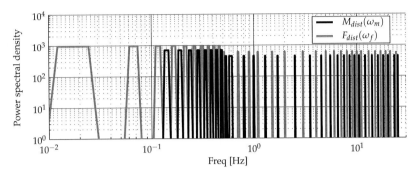

Figure 2.3: Power spectral density plot of disturbance signals F_{dist} and M_{dist}.

2.3.3 Design

The technique used to create the force disturbance signal $F_{dist}(t)$ and the motion disturbance signal $M_{dist}(t)$ was very similar. Both disturbance signals were multi-sines, defined in the frequency domain. First, a selection was made for the desired frequency range. To obtain a full bandwidth estimate of the admittance, a range between 0.01 Hz and 24 Hz was selected for the force disturbance signal $F_{dist}(t)$, which is sufficient to capture all arm dynamics [Perreault et al., 2001]. For the motion disturbance signal $M_{dist}(t)$, a range between 0.15 and 25 Hz was selected (the lower bound was selected such not to exceed the available motion space of the motion simulator).

For both disturbance signals, 31 logarithmically spaced frequency points were selected in the frequency range, such that two completely separated disturbance signals were obtained. Fig. 2.3 shows the spectral densities of the two disturbance signals. From the figure it can be seen that the signals are separated in the frequency domain, i.e., contained power at a different set of frequencies. To allow for frequency averaging, power was applied to two adjacent frequency points for each point [Schouten et al., 2008a], yielding 31 pairs of frequency points for each disturbance signal. The $F_{dist}(t)$ signal contained full power from 0.01 Hz up to 0.5 Hz and additional reduced power for the remaining frequencies. The $M_{dist}(t)$

signal contained full power from 0.15 Hz up to 0.5 Hz and additional reduced power for the remaining frequencies.

The phase of the sine components was randomized in order to obtain an unpredictable signal. A cresting technique was used to prevent large peaks in the time domain. By minimizing the crest factor (the compression or compactness of the signal), the power of the signal could be increased without changing signal amplitude [Pintelon and Schoukens, 2001; van der Ouderaa et al., 1988]. The crest factor is defined as the maximal amplitude of the signal divided by the root-mean-square (RMS) of the signal:

$$CF = \frac{\max |d(t)|}{RMS(d(t))} \qquad \text{2.4}$$

in which CF is the crest factor.

Using the inverse fast Fourier transform, time-domain signals $F_{dist}(t)$ and $M_{dist}(t)$ were obtained. These were cut to a length of 87 seconds. Both disturbance signals were smoothed using a fade-in and fade-out period, each 2 seconds, to minimize transient effects.

Before the motion disturbance signal $M_{dist}(t)$ could be used as the acceleration command signal for the simulator motion base, it needed to be corrected for divergent effects in velocity and position that would otherwise occur after integration. This was done by adding appropriate corrective signals to the fade-in period of the acceleration signals, resulting in a zero-mean velocity and zero-mean position deviations of the simulator motion base. The time-domain signals of the disturbances are shown in Fig. 2.4.

In the analysis of the measurement results, the fade-in and fade-out sections were removed from the measured signal, leaving 83 seconds of data. As data was recorded at 100 Hz, 8300 data samples were available for analysis, from which a set of 8192 ($= 2^{13}$) samples was selected.

2.4   Experiment

In the following it is described how BDFT dynamics – more specifically, B2P dynamics – and admittance can be measured simultaneously in an experiment. The results obtained from this experiment

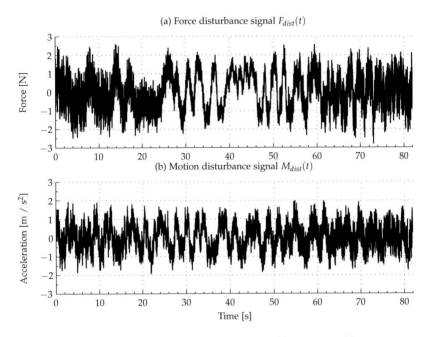

Figure 2.4: Disturbance signals $F_{dist}(t)$ and $M_{dist}(t)$.

served to validate the approach.

The *approach* presented here has been also used for the other experiments mentioned throughout this thesis. The *results* presented here are unique to the current chapter, as the experiment was the first of its kind and executed for only a limited number of participants. The data presented in other chapters were obtained through similar experiments of which the details will be provided in each chapter.

2.4.1 Hypotheses

There were two central hypotheses in this study. First of all, it was hypothesized that by using the disturbance signals $F_{dist}(t)$ and $M_{dist}(t)$, it would be possible to measure neuromuscular admittance and biodynamic feedthrough simultaneously. This hypothesis was tested in the following ways:

- Successful estimation of the admittance can be assumed when the admittance is comparable with results found in previous admittance experiments.

- Successful estimation of the biodynamic feedthrough, for which no data for comparison is available, can be assumed when the BDFT estimates show a high squared coherence (to be introduced in the following) and the estimates between different subjects show comparable shape and features.

- The contribution of the two disturbance signals should be distinguishable in the measured response, both from each other and from the residual noise in the measured response.

The second central hypothesis of this study was that a dependency of biodynamic feedthrough on the settings of the neuromuscular system exists. This dependency is proven if different biodynamic feedthrough dynamics are obtained for several control tasks for which it is known that they elicit changes in the neuromuscular admittance.

Regarding the expected results, one could argue that it can be expected that the 'stiff' behavior during the PT results in a strong mechanical coupling between the accelerations and the control device (due to the tight grip dynamics) and this will result in a stronger, more 'direct', feedthrough of accelerations.

One might also argue, however, that a stiffer neuromuscular system will reduce the feedthrough of accelerations, as a general property of stiff dynamical systems is it's capability to reject and attenuate disturbances.

Prior to the experiment it was unknown how the BDFT dynamics would exactly depend on the control task, that is, it was unknown how variations in neuromuscular admittance would influence the BDFT dynamics.

2.4.2 Apparatus

The experiment was performed on the SIMONA Research Simulator (SRS) of Delft University of Technology [Stroosma et al., 2003].

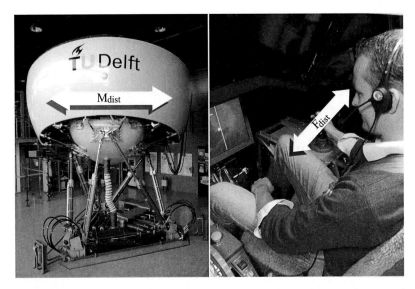

Figure 2.5: The experimental setup used in this study: a subject (right), seated inside the SIMONA Research Simulator (left) uses a side-stick to perform a control task. A force disturbance signal, $F_{dist}(t)$, perturbs the side-stick in lateral direction. A motion disturbance signal, $M_{dist}(t)$, perturbs the motion-base of the simulator in lateral direction.

The SRS is a six degrees-of-freedom research flight simulator, with a hydraulic hexapod motion system. The control device was an electrically actuated side-stick. No armrest for the arm that controlled the side-stick was present.

During the experiment the motion disturbance $M_{dist}(t)$ was applied to the simulator's motion base, the force disturbance $F_{dist}(t)$ was applied to the side-stick (see Fig. 2.5). The stick was fixed in longitudinal (pitch) direction, leaving the lateral (roll) direction free for control. A head-down display (15" LCD, 1024×768 pixels, 60 Hz refresh rate) was located in front of the right-hand pilot seat where subjects were seated during the experiment. The seat had a 5-point safety belt that was adjusted tightly in the experiments, to reduce torso motions. For the simulator motion, no motion filter was used,

Table 2.1: Subject data.

Subject	Height [cm]	Weight [kg]	Age [yrs]	Gender
1	162	55	25	Female
2	192	90	22	Male
3	178	73	19	Male
4	185	75	27	Male
5	169	63	23	Female
mean	177.2	71.2	23.2	
st. dev.	12.0	13.4	3.03	

meaning that the commanded acceleration signal was fed to the simulator motion base without filtering or washout.

2.4.3 Subjects

Five subjects (see Table 2.1) participated in this study. Subjects were recruited from the student population of Delft University of Technology. Their participation was voluntary, no financial compensation was offered.

2.4.4 Task and task instruction

During the experiment the side-stick was perturbed with the force disturbance signal $F_{dist}(t)$. The subjects executed the following disturbance rejections tasks (also referred to as the classical tasks):

- The position task (PT) (or stiff task), in which the instruction was to keep the position of the side-stick in the centered position, that is, to "resist the force perturbations as much as possible".

- The force task (FT) (or compliant task), in which the instruction was to minimize the force applied to the side-stick, that is, to "yield to the force perturbations as much as possible".

- The relax task (RT), in which the instruction was to relax the arm while holding the side-stick, that is, to "passively yield to all perturbations".

The human operator needed to set his/her neuromuscular properties differently for optimal control of each of the three control tasks. The PT is a task for which the best performance is achieved by being very stiff (i.e., a small admittance), the FT requires the operator to be very compliant (i.e., a large admittance). The RT is intended to yield an admittance which gives an indication of the passive dynamics of the neuromuscular system. An important difference between the RT and FT is the amount of neuromuscular activity involved. The RT asks for fully passive behavior with as little activity as possible. An alternative way of phrasing the RT instruction is to 'hang the weight of the arm on the stick'. The FT asks for maximum compliance, which requires a degree of neuromuscular activity to minimize the forces applied to the control device. The control device is being perturbed by an external force disturbance and the subject's task is to give way to these forces. An alternative way of phrasing the FT instruction is to 'follow the motions of the stick to the best of your abilities'.

The classical tasks have been used in numerous studies to investigate the variability of the neuromuscular system, e.g., [Abbink, 2006; Lasschuit et al., 2008; Mugge et al., 2009]. These same control tasks will also be used in the other experiments performed in the context of this thesis.

An important feature of the classical tasks is that the contribution of cognitive control inputs of the operator is only limited, just as was assumed when the scope of the research was defined (see Section 2.2.3). The control tasks are aimed at helping (or forcing) the human operator to attain three different levels of neuromuscular admittance, while minimizing the cognitive input.

Many of the disturbance frequencies are outside the area where visual control is possible (i.e., many of the disturbance frequencies are well above 1 Hz), so optimal performance is realized through using a particular setting of the neuromuscular system (e.g., by muscle co-contraction and the use of reflexes) and not by cognitive control actions based on, e.g., visual feedback. Occasional (abrupt) cognitive corrections that may occur are assumed to be of negligible influence as the transfer dynamics that are calculated from the data reflect the time average behavior during the experiment duration of 83 seconds.

During the PT and FT information was displayed on a screen in front of the subjects. A laterally moving red block displayed the parameter to be controlled and the reference was shown by a white vertical line running down the center of the display. During the PT the controlled parameter was the lateral side-stick deflection angle and the reference was 0 degrees; during the FT the controlled parameter was the applied force to the side-stick and the reference was 0 N. For both tasks the display also showed a time-history of position or force respectively such that subjects could monitor their performance. During the RT the display presented no information. Before entering the simulator the subjects were instructed on the goal of the experiment and the control tasks they were to perform. In the simulator, the subjects were seated in an aircraft seat with a side-stick positioned to the right. First, several training runs were performed to allow the subject to get used to the force and motion disturbances and the different control tasks. When the subjects indicated to have understood the control tasks the measurements started. Before the start of each experiment run the subject was asked to center the control stick to ensure the initial position of the involved limbs was similar for all experiment runs.

2.4.5 Disturbance signal scaling

During the experiment, the force disturbance signal $F_{dist}(t)$ was scaled in amplitude in an attempt to keep the standard deviations of the resulting control device deflections similar for all tasks. It is important to ensure that the same operating point of the neuromuscular system is excited for all control tasks, avoiding amplitude non-linearities and allowing to apply linear analysis methods to the intrinsically highly non-linear neuromuscular system [Cathers et al., 1999]. The gain used for the FT was 1, for the RT 2, and for the PT 8.

The original motion disturbance signal (see Fig. 2.4) contained a maximum acceleration of 2.46 m/s^2, a maximum velocity of 1.10 m/s and a maximum position deviation from the centered position of 0.90 m. For safety reasons it was chosen to stay well within the motion limits of the simulator (1.0 m to each side). It was chosen to use a gain factor of 0.8 on the motion disturbance signal. This yielded a maximum acceleration of 1.97 m/s^2, a maximum velocity of 0.88 m/s and a maximum position deviation from the centered position of 0.72 m. This gain was not varied over the experiment runs.

2.4.6 Independent variables

Two independent variables were used: the control task (TSK) and the motion disturbance signal (DIST). The different levels for each independent variable were:

- TSK: position task (PT); force task (FT); and relax task (RT)

- DIST: on; and off

Together, TSK and DIST yielded six different conditions. During the experiment, first the conditions with motion disturbance (DIST on) were executed in the order PT-RT-FT. Four repetitions of this sequence were executed. Then, the conditions without motion disturbance (DIST off) were executed, using the same control task order, PT-RT-FT, and three repetitions. For this preliminary study it was assumed that the influence of learning effects was negligible.

2.4.7 Dependent measures

During the experiments the angular deflection of the side-stick $\theta_{cd}(t)$ and the forces applied to the side-stick $F_{app}(t)$ were measured.

Calculating admittance and biodynamic feedthrough

The variables measured during the experiment, $\theta_{cd}(t)$ and $F_{app}(t)$, were averaged in the time domain over all available repetitions to reduce the noise in the signals [Schouten et al., 2008a]. The signals were then transformed to the frequency domain using the Fast Fourier Transform (FFT). The neuromuscular system dynamics are embedded in a closed-loop configuration (see Fig. 2.2), therefore a closed-loop identification method was adopted to estimate admittance, which will be briefly summarized here (see Chapter 3 for details).

The two disturbance signals, $F_{dist}(t)$ and $M_{dist}(t)$, are independent external inputs that affect the stick deflection $\theta_{cd}(t)$ and the applied force $F_{app}(t)$, which are dependent signals inside the closed-loop. The FRF estimate of the admittance, $\hat{H}_{adm}(j\omega_f)$, was calculated using the estimated cross-spectral density between $F_{dist}(t)$ and $\theta_{cd}(t)$ ($\hat{S}_{fdist,\theta}(j\omega_f)$) and the estimated cross-spectral density between $F_{dist}(t)$ and $F_{app}(t)$ ($\hat{S}_{fdist,f}(j\omega_f)$) [van der Helm et al., 2002]:

$$\hat{H}_{adm}(j\omega_f) = \frac{\hat{S}_{fdist,\theta}(j\omega_f)}{\hat{S}_{fdist,f}(j\omega_f)}$$

2.5

with ω_f the frequency range of the force disturbance signal $F_{dist}(t)$ and j the imaginary unit. $\hat{H}_{adm}(j\omega_f)$ is an estimate of the arm admittance $H_{adm}(j\omega)$ on frequency range ω_f.

As a measure for reliability of the estimate, the squared coherence $\hat{\Gamma}^2_{adm}(j\omega_f)$ was calculated [van der Helm et al., 2002]:

$$\hat{\Gamma}^2_{adm}(j\omega_f) = \frac{\left|\hat{S}_{fdist,\theta}(j\omega_f)\right|^2}{\hat{S}_{fdist,fdist}(j\omega_f)\hat{S}_{\theta,\theta}(j\omega_f)}$$

2.6

$\hat{\Gamma}^2_{adm}(j\omega_f)$ is a measure for the signal to noise ratio and thus for the linearity of the dynamic process. This function equals one when there are no non-linearities and no time-varying behavior and zero when there is no linear behavior at all.

In a similar way the transfer function describing the B2P dynamics, $H_{B2P}(j\omega_m)$, can be estimated, taking the motion disturbance $M_{dist}(t)$ as input and the control device deflection $\theta_{cd}(t)$ as output. The estimate of the B2P dynamics, $\hat{H}_{B2P}(j\omega_m)$, is calculated using the estimated cross-spectral density between $M_{dist}(t)$ and $\theta_{cd}(t)$

$(\hat{S}_{mdist,\theta}(j\omega_m))$ and the estimated auto-spectral density of $M_{dist}(t)$ $(\hat{S}_{mdist,mdist}(j\omega_m))$

$$\hat{H}_{B2P}(j\omega_m) = \frac{\hat{S}_{mdist,\theta}(j\omega_m)}{\hat{S}_{mdist,mdist}(j\omega_m)} \qquad \text{2.7}$$

with ω_m the frequency range of the motion disturbance signal $M_{dist}(t)$. The squared coherence function in this case is:

$$\hat{\Gamma}^2_{B2P}(j\omega_m) = \frac{\left|\hat{S}_{mdist,\theta}(j\omega_m)\right|^2}{\hat{S}_{mdist,mdist}(j\omega_m)\hat{S}_{\theta,\theta}(j\omega_m)} \qquad \text{2.8}$$

The use of periodic multi-sines as disturbance signals ensures that all spectral estimators are unbiased and have a relatively low variance [Pintelon and Schoukens, 2001].

Frequency decomposition

Both the measured total control device deflection $\theta_{cd}(t)$ and the applied force $F_{app}(t)$ can be separated into three parts: a contribution of the force disturbance $F_{dist}(t)$, a contribution of the motion disturbance $M_{dist}(t)$ and a residual contribution, which is not related to any disturbance signal, as shown in Eqs. 2.1 and 2.2. To obtain these separate contributions, a frequency decomposition technique was used. This technique will also be used in Chapters 4, 6 and 8 Frequency decomposition allows to separate the total measured signal into several contributions by splitting the total power spectral density function $PSD^{tot}(j\omega)$ of a signal by frequency. From $PSD^{tot}(j\omega)$, the contribution of the force disturbance signal $F_{dist}(t)$ (with frequency range ω_f) can be estimated by creating a PSD function - $PSD^{Fdist}(j\omega)$ - as follows:

$$PSD^{Fdist}(j\omega) = \begin{cases} PSD^{tot}(j\omega) & \text{for} \quad \omega = \omega_f \\ 0 & \text{for} \quad \omega \neq \omega_f \end{cases} \qquad \text{2.9}$$

That is, $PSD^{Fdist}(j\omega)$ is equal to $PSD^{tot}(j\omega)$ for the frequencies where $F_{dist}(t)$ contained power and $PSD^{Fdist}(j\omega)$ is zero for all other frequencies. The same can be done for the motion disturbance signal $M_{dist}(t)$, obtaining $PSD^{Mdist}(j\omega)$. For the remaining frequencies (where no disturbances were added) a final PSD function can be created, $PSD^{res}(j\omega)$, containing the residual power. It holds that:

$$PSD^{tot}(j\omega) = PSD^{Fdist}(j\omega) + PSD^{Mdist}(j\omega) + PSD^{res}(j\omega) \quad \boxed{2.10}$$

By using the inverse FFT, time-domain signals can be constructed from the different PSDs, obtaining the contribution of the disturbances and the residual in the time domain.

Frequency decomposition provides insight in the contribution (i.e., the 'effect') of the two disturbance signals to the total signal. The results can be used to confirm whether the disturbance signals excited the arm sufficiently to obtain reliable estimates for the admittance and B2P dynamics.

A requirement for a good estimate is that the response to the two disturbance signals can be identified separately in the total measured response. That is, the contribution of the two disturbance signals should be distinguishable, both from each other and from the residual part of the signal. This can be checked by investigating the PSD plots obtained by frequency decomposition.

Furthermore, using the time domain plots, the effect of the disturbance signals can be investigated in a quantitative manner, e.g., by calculating the standard deviation (SD) of the time responses the contributions of the disturbance signals can be compared in magnitude. Comparing the SD of the response for different tasks provides insight in the effect of scaling the force disturbance signal, as discussed in Section 2.4.5.

2.5   Results

2.5.1   Admittance and biodynamic feedthrough

Fig. 2.6 shows admittance estimates of a typical subject, obtained from the analysis of data measured when both force and motion disturbances were applied. As expected, for low frequencies the admittance is the largest for the force task and the smallest for the position task. At higher frequencies the differences become smaller

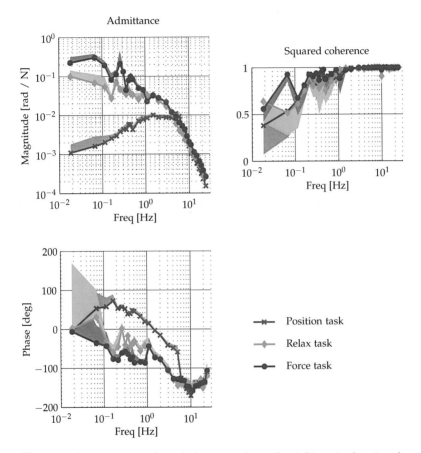

Figure 2.6: Neuromuscular admittance estimate for Subject 5, showing the mean (lines) and standard deviations (colored bands) of the estimates over 4 repetitions of each task. The figure shows the magnitude, the phase and the squared coherence. For the magnitude and phase mean + 1 SD is shown, for the squared coherence mean - SD is shown.

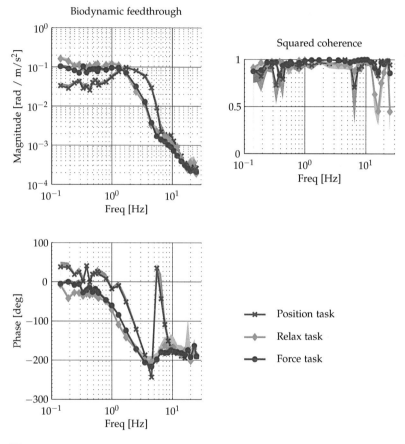

Figure 2.7: B2P estimate for Subject 5, showing the mean (lines) and standard deviations (colored bands) of the estimates over 4 repetitions of each task. The figure shows the magnitude, the phase and the squared coherence. For the magnitude and phase mean + 1 SD is shown, for the squared coherence mean - SD is shown.

as dynamics are more and more governed by inertia. The admittance measured for the relax task was expected to lie between the one measured for the force task and the position task. However, for several frequencies the magnitude of the admittance measured during RT is hardly distinguishable from the one measured during the FT. This was observed for several subjects.

The lack of difference between RT and FT could be explained by the fact that the force gain scaling factor of the relax tasks was set too high, yielding too large control device deflections relative to the two other tasks (more on this in the next section). Another possible factor is that, although some time for training was scheduled, some subjects indicated to have difficulty distinguishing between tasks, especially between the FT and the RT.

High squared coherences, a measure for reliability of the estimate, were found at all frequencies for all tasks, except for the lowest frequency. Furthermore, the results were found to be comparable with the results of previous studies [Abbink, 2006; Lasschuit et al., 2008; Mugge et al., 2009]. The differences observed between control tasks are caused by adaptations of the neuromuscular system.

Fig. 2.7 shows B2P estimates for a typical subject, measured simultaneously with the admittance shown in Fig. 2.6. Also here, the differences in B2P for the different control tasks are caused by adaptations of the neuromuscular system by the human operator in response to task instruction. It can be readily seen that the B2P depends on task instruction, and thus on the neuromuscular admittance. The reliability of the measurement is reflected in the high squared coherences that were obtained for all frequencies.

The B2P dynamics measured for the other subjects showed comparable shapes and features. This can be clearly demonstrated by averaging over all subjects. Fig. 2.8 shows the mean and standard deviation of the B2P magnitude for the three control tasks, averaged over all subjects. The figure shows a clear dependency of B2P on task instruction and hence the setting of the neuromuscular system. For low frequencies, the B2P is the lowest for the PT and the highest for the RT.

Surprisingly, for frequencies higher than approximately 1.5 Hz, the B2P of the PT is higher than the one found for the RT. Moreover,

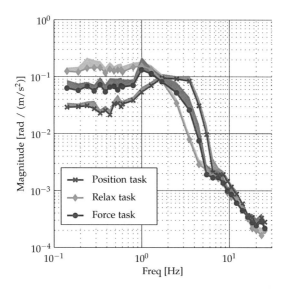

Figure 2.8: Mean (lines) and standard deviation (colored bands, mean + 1 SD) of the B2P magnitude, averaged over all subjects, for all control tasks.

a peak in B2P is observed for the PT between approximately 2-3 Hz. This result is remarkable and suggest that the strategy attained during the PT (being 'stiff'), leads to an increase in the feedthrough of motion disturbances above 1.5 Hz, in comparison to the other control tasks [Venrooij et al., 2009].

2.5.2 Frequency decomposition

Fig. 2.9 shows an example of a plot that results from frequency decomposition. Such a plot is referred to as a power spectral density (PSD) plot. This example shows the PSD of the measured control force, $F_{app}(t)$. In this figure, both the total PSD and the contributions are shown of the force disturbance $F_{dist}(t)$, the motion disturbance $M_{dist}(t)$ and the residual. It can readily be observed that far more power was measured at the frequencies where disturbances were added than at other frequencies. In fact, all peaks in the power spectral density plot can be attributed to a disturbance, either the

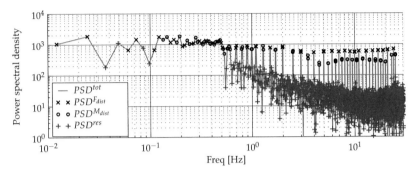

Figure 2.9: Result of the frequency decomposition of the measured applied force $F_{app}(t)$ for the relax task in the frequency domain (Subject 5).

force disturbance or the motion disturbance. From this, it can be concluded that both disturbances excited the arm of the subject sufficiently to result in a measurable response.

Furthermore, the figure shows that the power measured at the disturbance frequencies is one to several orders higher than the power measured at the frequencies where no disturbance was added. This indicates the responses to the disturbance signals are predominantly found at the same frequencies as disturbances itself. In other words, there is no or little 'leakage' to other frequencies than the disturbance frequencies. These observations lead to the conclusion that the contribution of the two disturbance signals is distinguishable, both from each other and the residual part of the signal. This is a requirement to obtain reliable estimates for the neuromuscular admittance and the B2P.

Fig. 2.10 shows the results of taking the inverse FFT of the PSDs in Fig. 2.9. The obtained plot is referred to as the time domain plot. In the figure, the total time response is shown in (a) and the three contributions in (b),(c) and (d). The sum of the three contributions yields the total signal again. Note that the residual shows a low frequency component. This component can also be identified in the PSD plot, by the relatively high power the residual, $PSD^{res}(j\omega)$, contains for low frequencies (<0.1 Hz). The low frequency component is an artifact caused by the low frequency force disturbance

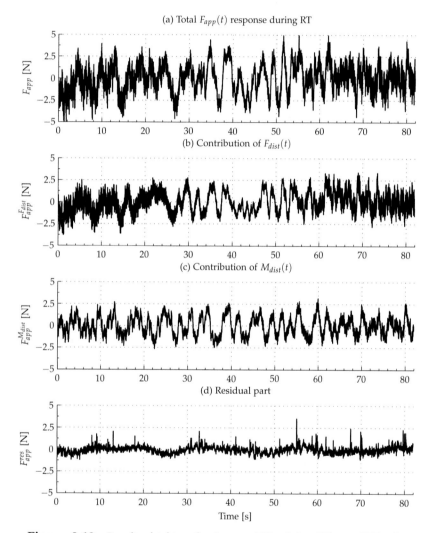

Figure 2.10: Result of taking the inverse FFT of the different PSDs from Fig. 2.9 ($F_{app}(t)$, RT, Subject 5), with: (a) the total measured signal, (b) contribution of the force disturbance F_{dist}, (c) contribution of the motion disturbance M_{dist} and (d) the residual part.

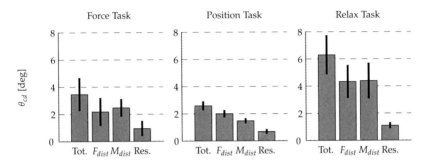

Figure 2.11: Mean and 95% confidence interval of the standard deviation of the total time-domain signal, the contributions of the two disturbances and the residual part for each task for the measured control device deflections $\theta_{cd}(t)$, averaged over all subjects.

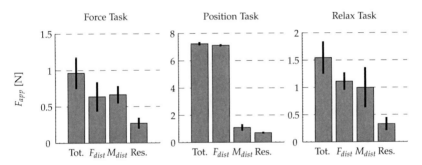

Figure 2.12: Mean and 95% confidence interval of the standard deviation of the total time-domain signal, the contributions of the two disturbances and the residual part for each task for the measured applied force $F_{app}(t)$, averaged over all subjects. Note the different scaling of the y-axis.

offered at neighboring frequencies. In other words, for very low frequencies, some 'leakage' does seem to occur.

By calculating the SD of the time domain plots of Fig. 2.10 the relative contributions can be compared in a more quantitative manner. Fig. 2.11 shows for each task the magnitude of the control device deflection $\theta_{cd}(t)$, averaged over all subjects, calculated by taking the SD of the time domain plots for each subject. The figure shows

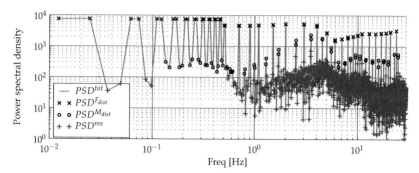

Figure 2.13: Result of the frequency decomposition of the measured applied force $F_{app}(t)$ for the position task in the frequency domain (Subject 5).

means across subjects and 95% confidence intervals. The data provide insight in the effect of scaling the force disturbance signal.

Recall that the force disturbance signal was scaled in an attempt to obtain similar control device deflections, to ensure linearity. When looking at the result for the total signal for each task, it can be concluded that the deflections measured for the FT (with scaling gain 1) and the PT (with scaling gain 8) were comparable in magnitude (around 3 degrees). However, the deflections obtained during the RT (with scaling gain 2) were higher than the deflections obtained for the other two tasks (around 6 degrees). From this it can be concluded that the scaling gain for the RT was too high. As a consequence, the admittance and B2P measured during the RT cannot be directly compared with the other tasks. It is highly likely that the scaling factors used in the experiment increased the measured admittance during the RT with respect to the one measured during the FT, explaining the lack of difference between the admittance measured for the two tasks. The experiment results could be improved by reducing the gain for the RT such that it yields similar control device deflections.

Furthermore, the figure shows that both the force disturbance and the motion disturbance significantly contribute to the measured control device deflection $\theta_{cd}(t)$ for all control tasks. The contribution of both disturbances is a number of times higher than the contribution of the residual, which is not related to any disturbance. The same can be concluded for the contributions measured for the applied force $F_{app}(t)$, shown in Fig. 2.12.

A notable exception is the position task. For this task, the contribution of the motion disturbance signal is much lower than the contribution of the force disturbance, and, in fact, of similar magnitude as the residual response. The explanation for this lies in the character of the position task, where the subject is requested to resist the force perturbations. Evidently, during such a task the motion disturbance contributes far less than the force disturbance in the measured applied force.

However, to be able to estimate the B2P reliably, the response to the motion disturbance signal should still be distinguishable from the residual response. That this is in fact the case can be seen in Fig. 2.13. This figure shows the PSD plot of the $F_{app}(t)$ for the position task for a typical subject. A clear increase in measured power is visible for the frequencies where the motion disturbance was added (indicated by the circles). This shows that also for this task the response to the motion disturbance can be distinguished from the residual part at every frequency, allowing for a reliable estimate of the B2P.

2.6 Conclusions

For the studied experimental conditions, it was concluded that the proposed method was successful in the simultaneous measurement of admittance and B2P, based on the following evidence:

- The two disturbances successfully excited the arm and each resulted in control inputs that are distinguishable, both from each other and the residual noise;

- The admittance measurements are comparable to the results found in other studies in which admittance was measured during side-stick control [Lasschuit et al., 2008]. The reported high coherences, signifying good signal-to-noise ratios, indicate that the admittance estimates are reliable;

- For the B2P measurements, high coherences were found, indicating the B2P estimates are reliable. Furthermore, between subjects, each task shows a B2P with comparable shape and features.

The experimental results expose a relationship between the B2P dynamics and neuromuscular admittance, as was already hypothesized. The differences that were observed in the B2P dynamics for the different control tasks can only be caused by the changes in the setting of the neuromuscular system. Hence, it can be concluded that there exists a dependency of B2P on neuromuscular admittance. As the neuromuscular system is highly adaptable, and its settings depend on factors like task instruction, this dependency is highly relevant when studying BDFT. The proposed measuring method allows measuring B2P and neuromuscular admittance simultaneously, providing insight in how exactly the settings of the neuromuscular system influence the level of BDFT.

These results can be used in three primary ways. The first is to increase the fundamental understanding of biodynamic feedthrough and its underlying processes. One way of increasing this knowledge is by modeling the interactions between admittance and BDFT, as is done in Chapter 4. The second is to gain knowledge on which neuromuscular setting is most effective in reducing BDFT. For example, the results in Fig. 2.8 show that a low admittance (relating to a PT) results in a low level of B2P for disturbances at frequencies below 1.5 Hz (but not above). Finally, these results can also be used to develop a solution to biodynamic feedthrough, in the form of a canceling controller. The dependency between BDFT and neuromuscular admittance shows that the correct canceling control action is dependent on the current setting of the neuromuscular system of the human operator. Note that an effective canceling controller should thus be able to detect and adapt to changes in the HO's neuromuscular system (see also Chapter 6).

CHAPTER 3

A framework for biodynamic feedthrough analysis

This chapter presents a framework for BDFT analysis. The framework contains definitions, nomenclature and dynamical relationships. In this chapter, it is shown how the dynamical relationships can be obtained from measurement and how they are interrelated to each other. By using experimental measurement data, the validity of the derived relationships is demonstrated and the practical applicability of the framework is illustrated. The results show that the proposed framework offers a versatile toolbox for analyzing and understanding BDFT problems.

The contents of this chapter are based on:

Paper title   A New View on Biodynamic Feedthrough Analysis: Unifying the Effects on Forces and Positions

Authors   Joost Venrooij, Mark Mulder, David A. Abbink, Marinus M. van Paassen, Frans C. T. van der Helm, Heinrich H. Bülthoff, and Max Mulder

Published in   IEEE Transactions on Cybernetics, Vol. 43, No. 1, Feb. 2013

- and -

Paper title   A Framework for Biodynamic Feedthrough Analysis – Part I: Theoretical Foundations

Authors   Joost Venrooij, Marinus M. van Paassen, Mark Mulder, David A. Abbink, Max Mulder, Frans C. T. van der Helm, and Heinrich H. Bülthoff

Published in   IEEE Transactions on Cybernetics, Vol. 44, No. 9, Sep. 2014

- and -

Paper title   A Framework for Biodynamic Feedthrough Analysis – Part II: Experimental Validation

Authors   Joost Venrooij, Marinus M. van Paassen, Mark Mulder, David A. Abbink, Max Mulder, Frans C. T. van der Helm, and Heinrich H. Bülthoff

Published in   IEEE Transactions on Cybernetics, Vol. 44, No. 9, Sep. 2014

3.1 Introduction

One of the main complexities of biodynamic feedthrough (BDFT) is that there is little consensus in the literature on how to approach the BDFT problem in terms of definitions, nomenclature and mathematical descriptions. Across studies many different identifiers have been used to describe BDFT phenomena, such as vibration breakthrough [Jex and Magdaleno, 1978; McLeod and Griffin, 1989], biodynamic coupling [Idan and Merhav, 1990] and operator-induced oscillations [Sirouspour and Salcudean, 2003] amongst other examples.

Although significant progress in understanding BDFT has been made over the past decades, this progress is derived from fragmented developments rather than from a unified research effort. The fragmentation of the BDFT research field is one of the reasons why no satisfactory answer has yet been formulated to important questions like: which role do the human body dynamics actually play in the occurrence of BDFT? What is the influence of individual components in the human-machine system on BDFT dynamics and how does the influence of several components combine? How can the effectiveness of a BDFT mitigation technique be explained or predicted? How to best mitigate the occurrence of BDFT while minimizing the impact on voluntary control inputs?

Efforts to unify the currently existing BDFT literature are obstructed by the diversity of works and complexity of the BDFT problem. A basic framework, consisting of a taxonomic model of human operator processes contributing to performance in vibration has been described by Lewis [1974] and Lewis and Griffin [1976]. At the time of its conception, the degree of predictive capabilities of the framework were limited due to the "many gaps in knowledge about the response of the various sub-systems to vibration" ([Lewis and Griffin, 1976], p. 203). Since that time, many additional insights have been gained which have allowed to close some of these gaps at least partially. However, the diversity between these studies does not readily allow for a synthesis of the results back into the proposed framework. For example, the fact that many similar phenomena and dynamics have received different names across studies, or, equally confusing, that the same name was used for different phenomena and dynamics, complicates a comparison. Currently, the impact of a given work or the relationship between two works often remains obscured, mainly due to the absence of a common framework.

This chapter aims to provide such a framework for biodynamic feedthrough analysis. Its theoretical foundations will be provided, accompanied by evidence for the framework's validity. In order to gain some intuition into what the different types of dynamics can tell us about BDFT, a practical interpretation of the experimentally obtained transfer functions is provided. Finally, the usefulness of the framework is demonstrated by applying it to experimental data and in a meta-analysis of existing literature.

The goal of this framework is twofold. First of all, it aims to provide a common ground to study, discuss and understand BDFT and its related problems. Simply said, it aims to add clarity to the field of BDFT by defining 'what is what'. Using this framework, old and new BDFT research can be (re)interpreted, evaluated and compared. Secondly, and equally important, the framework itself allows for gaining novel insights into the BDFT phenomenon. The mathematical relationships that are included in the framework allow for a better understanding regarding the influence of individual components on BDFT dynamics.

What the current framework adds to the already quite extensive body of knowledge on BDFT is a comprehensive set of definitions, nomenclature and mathematical notations. In addition, from the relationships between different types of BDFT dynamics new insights emerge, such as how the control device dynamics influence biodynamic feedthrough dynamics, or how different mitigation techniques can be represented in a unified way.

The framework proposed here will be used in the remaining chapters of this thesis, where it serves to disambiguate which signals were measured, which dynamics were modeled and which method of BDFT mitigation was used. In doing so, some of the unanswered questions regarding BDFT can be answered, but many questions will need further research. Therefore, the framework should also be used in future studies. In order to incorporate novel insights, the framework is open to additions and extensions, paving the way to unify the efforts of individual research groups investigating BDFT related phenomena.

The structure of this chapter is as follows: first, in Section 3.2 and Section 3.3, definitions and nomenclature will be proposed and the role of different signals and dynamics in the occurrence of BDFT will be addressed. In Section 3.4 it will be shown how these signals and dynamics can be obtained from measurement. This is followed by a discussion on the interrelation of different dynamics in Section 3.5 and Section 3.6. By using experimental measurement data, the validity of the derived relationships is demonstrated in Section 3.7. Then, by looking at the experimentally obtained transfer functions an intuition for the different dynamics is gained in Section 3.8. In Section 3.9, the goal of the proposed framework to increase our understanding of BDFT is illustrated through a practical example. In Section 3.10, the descriptive capabilities of the framework are illustrated using three selected works from literature. Finally, the conclusions are presented in Section 3.11.

3.2 The BDFT system model

Fig. 3.1 shows the biodynamic feedthrough system model, which was already briefly mentioned in Chapter 2. In this chapter, the model is revisited in full detail.

The biodynamic feedthrough system model is a conceptual model and shows all elements of a typical biodynamic feedthrough system. Each model block contains a transfer function (indicated with H) describing the dynamics of the system it represents. Tables 3.1 and 3.2 contain a brief description of the elements and the signals present in the BDFT system model respectively.

First, let's consider a human-machine system without the influence of acceleration disturbances. Such a system consists of a human operator (HO) and a controlled element (CE), e.g., a vehicle. The HO is manually controlling the CE using a control device (CD). The HO generates control commands by comparing the current state of the CE $y_{cur}(t)$ with a goal state $y_{goal}(t)$. Based on differences

Table 3.1: BDFT system model elements.

Element	Description
H_{CD}	Control device dynamics
H_{CDFT}	Control device feedthrough dynamics; effect of M_{dist} on CD
H_{CE}	Controlled element dynamics; system under control by HO
H_{CNS}	Central nervous system dynamics; brain and spinal cord of HO
H_{HO}	Human operator dynamics
H_{HOCD}	Interface dynamics between HO and CD; e.g., grip dynamics, armrest
H_{NMS}	Neuromuscular system dynamics; muscles, bones, etc., of HO
H_{PLF}	Platform dynamics; source of motion disturbance M_{dist}
H_{PLFHO}	Interface dynamics between PLF and HO; e.g., seat dynamics

Table 3.2: BDFT system model signals.

Signal	Description
$\theta_{cd}(t)$	Control device deflection (position)
$d(t)$	External disturbance on CE
$F_{app}(t)$	Force applied on control device (externally)
$F_{arm}(t)$	Force applied by the human operator (here, through arm)
$F_{arm}^{rem}(t)$	Force applied as result of operator remnant
$F_{cdft}(t)$	Control device feedthrough force
$F_{dist}(t)$	Force disturbance on control device
$F_{tot}(t)$	Total force on control device
$M_{dist}(t)$	Motion disturbance, originating from PLF
$n_{cog}(t)$	Cognitive (voluntary) control signal (neural commands)
$y_{cur}(t)$	Current state of CE
$y_{err}(t)$	Difference between goal and current state of CE
$y_{goal}(t)$	Goal state of CE

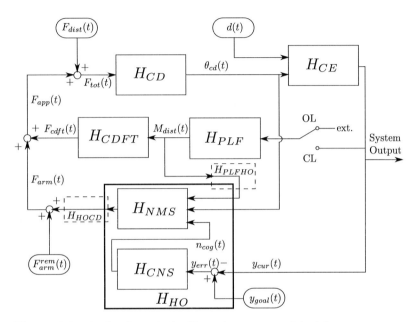

Figure 3.1: The biodynamic feedthrough system model. A human operator (HO) controls a controlled element (CE) using a control device (CD). Motion disturbances $M_{dist}(t)$ are coming from the platform (PLF). The feedthrough of $M_{dist}(t)$ to involuntary applied forces $F_{arm}(t)$ and involuntary control device deflections $\theta_{cd}(t)$ is called biodynamic feedthrough (BDFT). The feedthrough of $M_{dist}(t)$ to inertia forces $F_{cdft}(t)$ is called control device feedthrough (CDFT). $F_{app}(t)$ is the sum of the forces applied to the control device by the HO. The HO consists of a central nervous system (CNS) and a neuromuscular system (NMS). The connection between the HO and the environment is governed by two 'interfaces', H_{PLFHO} and H_{HOCD}. The CE and PLF can form an open-loop (OL) or closed-loop (CL) system. The focus of the current chapter: a framework for BDFT analysis is introduced, in which the BDFT system model plays a central role.

between these two states $y_{err}(t)$, the HO's central nervous system (CNS) – which is responsible for all cognitive control commands – formulates a voluntary control command, described here as a cognitive supra-spinal input $n_{cog}(t)$, which is transmitted neurally to the neuromuscular system (NMS). The NMS represents the dynamics of the limb connected to the CD and contains body parts such as bones, muscles, etc. The CNS includes the corticospinal tract or 'upper motor neurons', the NMS includes the spinal tract or 'lower motor neurons'. The NMS, which in this thesis is assumed to be a human arm, exerts a force $F_{arm}(t)$ on the CD. Note that also other forces may act on the CD, such as haptic feedback forces or external force perturbations. The control device deflections $\theta_{cd}(t)$, form the control input for the CE. Note that the CE can also be perturbed by a disturbance signal $d(t)$, for which the HO should compensate.

The representation of the manually controlled human-machine system can be extended to account for the effect of accelerations. These typically originate from the motion of a vehicle, referred to as the platform (PLF). The acceleration signal, coming from H_{PLF}, is called the motion disturbance signal $M_{dist}(t)$. The influence of the motion disturbance signal on the human-machine system is modeled through two effects: first, the mass of the control device m_{cd} converts the PLF accelerations into inertial forces (also known as fictitious forces or d'Alembert forces) $F_{cdft}(t)$. This effect is described in the H_{CDFT} block and is labeled here control device feedthrough (CDFT). Second, the PLF accelerations are transferred through the body of the HO and induce unintentional motions in the limb that is in contact with the CD, thereby leading to unintentional forces applied to the control device and – if the control device is movable, i.e., not rigid – these result in involuntary deflections of the control device. The generation of both involuntary forces and involuntary deflections is what is defined here as biodynamic feedthrough (BDFT):

Biodynamic feedthrough

The transfer of accelerations through the human body during the execution of a manual control task, causing involuntary forces being applied to the control device, which may result in involuntary control device deflections.

For the control device feedthrough (CDFT) the following definition is proposed:

Control device feedthrough

The transfer of accelerations through the control device mass, resulting in inertial forces being applied to the control device.

It should be noted that acceleration disturbances influence the human body in various other ways as well [Lewis and Griffin, 1976; McLeod and Griffin, 1989]. Examples of these biodynamic interferences are the occurrence of visual impairment (blurring of visual perception) and neuromuscular interference (reducing the signal-to-noise ratio between voluntary and involuntary muscle activity). These are not included in the BDFT system model presented here, which is only concerned with modeling the effects of BDFT.

BDFT manifests itself as an additional involuntary component in $F_{arm}(t)$. As can be observed in Fig. 3.1, $F_{arm}(t)$ is comprised of four components, of which three are related to the signals entering the H_{NMS} block. One of these components is due to motion disturbance $M_{dist}(t)$. The fourth component is the remnant, $F_{arm}^{rem}(t)$, which is also indicated in Fig. 3.1. Remnant can be defined as the operator's control output power that is not linearly correlated with the system inputs (such as the forcing functions) [McRuer and Jex, 1967].

The effects of BDFT are 'transferred' from $F_{arm}(t)$ into $F_{app}(t)$, into $F_{tot}(t)$, and into $\theta_{cd}(t)$. In that way, the control device deflections will have an involuntary component, which are BDFT induced involuntary control device deflections.

When the operator is both on board *and* controlling the vehicle, a connection exists between the CE and the PLF, as the HO's inputs affect the PLF's motion. This situation is referred to as a closed-loop (CL) BDFT system, a type of BDFT system that can lead to weakly damped or unstable oscillations. The alternative situation, an open-loop (OL) BDFT system, occurs if the HO is a passenger on board of a moving vehicle and engaged in a manual control task other than control of that same vehicle. Both closed-loop and open-loop BDFT systems are important and practically relevant. In the model, these two types are included through a switch which can either open or close the loop between CE and PLF. In the open-loop case, the PLF receives inputs from outside the human-machine system considered here. These external inputs are indicated in Fig. 3.1 as 'ext.'

The two dashed boxes shown in Fig. 3.1 are the two 'interfaces' that govern the connection between the operator and the environment. These blocks are indicated with dashed lines as they do not contain the dynamics of a single physical element, but dynamics that are influenced by several other systems. In the analysis the interface dynamics are therefore often lumped with other dynamics. H_{PLFHO} describes the dynamics of the connection between the PLF and the HO, influenced by, for example, seat suspension or seat belts, representing the interaction dynamics with the body of the HO. These dynamics are sometimes referred to as the 'seat transmissibility' and they determine how accelerations enter the operator's body. The interface H_{HOCD} describes the dynamics of the connection between HO and CD, e.g., grip visco-elasticity or the effect of an armrest. This interface determines how limb motions result in – voluntary and involuntary – forces $F_{arm}(t)$.

Finally, Fig. 3.1 shows the addition of a force disturbance signal, $F_{dist}(t)$. This force signal is applied to the control device and is used to obtain an estimate of the dynamics of the human limb in contact with the CD. The results presented in Chapter 2 showed that BDFT is strongly dependent on these limb dynamics. To describe the adaptive dynamics of human limbs the neuromuscular admittance is used. Admittance can be defined as [Abbink et al., 2011]:

Neuromuscular admittance

The causal dynamic relationship between the force acting on the limb (input) and the position of the limb (output).

The admittance, when determined by linear time-invariant (LTI) estimation techniques, shows properties of a mass-spring-damper system due to visco-elastic properties of the muscle and the limb inertia, as well as higher-order dynamics due to reflexive activity and grip dynamics. Some of the physical properties underlying the admittance can be assumed to be time-invariant, such as inertia, reflexive time delays, but others are highly adaptive such as the reflexive activity and muscle co-contraction.

3.3 BDFT signals and dynamics

This section provides a closer look at the different BDFT system elements, signals and dynamics introduced previously. In the following, the control device will be assumed to be rotational, of which the deflections are expressed here in [rad] and the PLF motion will be assumed to be linear, here expressed in [m/s^2]. However, the discussion generalizes also to linear control devices (with deflections in [m]) or rotational PLF motions (with accelerations in [rad/s^2]). The force disturbance will be expressed in [N] here, but could also be expressed in other units, e.g. [Nm].

3.3.1 Signals

Many of the signals shown in Fig. 3.1 can be measured directly in an experimental environment, but also in actual vehicles. Control device deflections $\theta_{cd}(t)$ [rad] and the forces applied to the control device $F_{app}(t)$ [N] can be measured using a control device equipped with appropriate sensors. Note that the control device feedthrough forces $F_{cdft}(t)$ [N] are technically *not* part of BDFT, as they do not feed through the human body. They are, however, present in the measured signal $F_{app}(t)$. How to deal with the CDFT forces will be addressed in a later section.

Force disturbance $F_{dist}(t)$ [N] can be assumed to be known. In an experimental environment, this signal is typically designed by the experimenter. Signal $F_{tot}(t)$ [N] is the sum of the measurable force $F_{app}(t)$ and the additional force disturbance $F_{dist}(t)$. Acceleration disturbance signal $M_{dist}(t)$ [m/s^2] can be obtained through an accelerometer or other forms of inertial measurement. In open-loop experiments, this signal can be designed by the experimenter and needs not to be measured, if the commanded disturbance signal is accurately followed by a motion base platform containing the HO and CD.

Of the remaining signals, it may be possible to obtain $d(t)$ and $y_{cur}(t)$, but for $y_{goal}(t)$ and $n_{cog}(t)$ this is more challenging, if possible at all. When studying BDFT, however, these signals are only of secondary interest. These signals are mostly relevant for the generation of *voluntary* control inputs. As BDFT is exclusively *involuntary* in nature, the voluntary control inputs are not a prime concern when measuring, analyzing and understanding BDFT at a fundamental level. It will be shown that BDFT can be studied in detail without requiring knowledge on the voluntary and/or cognitive elements. The force applied as result of operator remnant $F_{arm}^{rem}(t)$ can be assumed to be small compared to the other contributions and is therefore ignored for the remainder of this discussion.

By following the path from $M_{dist}(t)$ to $F_{arm}(t)$ and $\theta_{cd}(t)$ in Fig. 3.1, it can be observed that this path does *not* include the systems H_{CE}, H_{PLF} and H_{CNS}. The first two of these blocks describe vehicle dynamics, which are relevant to describe the acceleration signals the HO is exposed to, and – in closed-loop systems – the vehicle motion in response to BDFT induced inputs. These blocks are therefore of interest when studying, e.g., the stability of the human-vehicle interaction in the presence of BDFT [Quaranta et al., 2012], however, they do not influence the BDFT dynamics directly. The H_{CNS} block is responsible for generating *cognitive* control inputs, i.e., voluntary control behavior. As biodynamic feedthrough is strictly involuntary in nature, the contents of this block are not influencing the occurrence of BDFT directly either. Indirectly, it is possible that some involuntary BDFT induced control inputs are cognitively corrected for by the pilot. As it is known that the cognitive bandwidth is

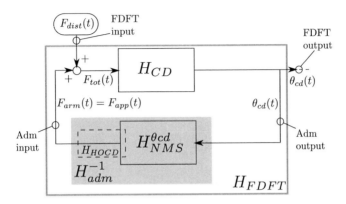

Figure 3.2: The dynamical elements that play a role in the response to force disturbances: control device dynamics, admittance and the force disturbance feedthrough (FDFT) dynamics.

usually limited to frequencies well below 1 Hz [Allen et al., 1973], it can be assumed that the CNS influences the BDFT dynamics primarily below this frequency [Quaranta et al., 2012]. In order to study the effects of, e.g., compensatory or feedforward control actions on the occurrence of BDFT, an operator model of choice can be included in the CNS block. If required, the current output of the CNS block $n_{cog}(t)$ can be changed to the output of the selected operator model and inserted at the correct location in the BDFT system model. More research into the influence of cognitive control actions on BDFT dynamics is certainly required, but is left here for future work.

3.3.2   Response to force disturbances

From the BDFT system model a range of dynamical relations can be derived. Before focusing on BDFT itself, first the responses of the control device and the neuromuscular system to *force disturbances* are presented. These dynamical relationships will play a large role in the later discussion on BDFT, which deals with the response to *motion disturbances*.

By removing the secondary systems and signals – H_{CE}, H_{CNS}, H_{PLF},

$d(t)$, $y_{cur}(t)$, $y_{goal}(t)$ and $n_{cog}(t)$ – from Fig. 3.1 and by removing the influence of vehicle motion – for now – by setting $M_{dist}(t)$ to zero, Fig. 3.2 is obtained, showing the control device H_{CD}, the interface H_{HOCD} and the neuromuscular system $H_{NMS}^{\theta cd}$. The superscript $^{\theta cd}$ in the latter signifies that the original multi-inputs-single-output (MISO) system H_{NMS} from Fig. 3.1 has been reduced to a single-input-single-output (SISO) system with $\theta_{cd}(t)$ as input.

Control device dynamics

The control device dynamics H_{CD} can typically be represented by a second-order mass-spring-damper (MSD) system, like [Venrooij et al., 2014a]:

$$H_{CD}(s) = \frac{1}{I_{cd}s^2 + b_{cd}s + k_{cd}}$$
$$= \frac{\theta_{cd}(s)}{F_{tot}(s)}$$

3.1

where:

- s is the Laplace variable,

- I_{cd} is the control device inertia [N m s^2 / rad],

- b_{cd} is the control device damping [N m s / rad],

- k_{cd} is the control device stiffness [N m / rad] and

- $\theta_{cd}(s)$ and $F_{tot}(s)$ are Laplace transforms of their respective time signals.

Admittance

The admittance includes the dynamics of the neuromuscular system and grip dynamics [Abbink, 2006], as indicated in the filled gray box in Fig. 3.2. It is important to note that, according to the definition, admittance describes the dynamics between forces $F_{arm}(t)$

(input) and positions $\theta_{cd}(t)$ (output), which is opposite to the directions of the arrows in Fig. 3.2. This implies that in the figure the *inverse* of admittance, H_{adm}^{-1}, is shown. The inverse of admittance is also known as impedance. The following holds:

$$H_{adm}^{-1}(s) = H_{NMS}^{\theta cd}(s)H_{HOCD}$$
$$= \frac{F_{arm}(s)}{\theta_{cd}(s)}$$

3.2

The admittance represents the position deviation of a limb in response to a force disturbance (in [rad / N]). A low/small admittance (high impedance) means a force disturbance elicits only a small position deviation, implying 'stiff' limb dynamics. Correspondingly, a high/large admittance (low impedance) implies 'compliant' limb dynamics.

Force disturbance feedthrough

The combined dynamics of the coupled system containing the control device and the human arm is described by the force disturbance feedthrough (FDFT) dynamics [Venrooij et al., 2010a, 2013a].

Force disturbance feedthrough

The transfer dynamics of force disturbances acting on an operator's limb, exciting the coupled systems of the human body and control device, resulting in control device deflections.

FDFT can be expressed mathematically as:

$$H_{FDFT}(s) = \frac{\theta_{cd}(s)}{F_{dist}(s)}$$

3.3

The FDFT dynamics are indicated in the outlined gray box in Fig. 3.2. The FDFT dynamics represent how external force disturbances result in control device deflections (in [rad / N]). A high FDFT means a force disturbance causes large limb and control device deflections. How admittance, control device and FDFT dynamics are related is addressed in Section 3.5.

3.3.3   Response to motion disturbances

Shifting focus from the response to force disturbances to the response to motion disturbances requires putting $F_{dist}(t)$ to zero and
re-inserting $M_{dist}(t)$, resulting in Fig. 3.3.

In the literature, the response to motion disturbances, i.e., the biodynamic feedthrough, has been measured in roughly two distinctive ways. In a study by Sövényi and Gillespie [2007], biodynamic
feedthrough was determined by measuring the forces that passive
human operators (i.e., operators not involved in any particular control task) applied to a fixed control device, in this case a side-stick,
while being subjected to an acceleration disturbance. The stick was
"fixed in its vertical position" using a peg, so only the involuntary
forces applied to the control device were measured. In the frequency
domain, the results showed common features between all subjects,
i.e., two peaks and a notch, but dynamics differed significantly between subjects. This was attributed to differences in anthropometric properties between subjects, variations in posture, differences
in grip and in nominal muscle activation, but was not investigated
further.

In other studies, e.g., in [Sirouspour and Salcudean, 2003] and [Ven-
rooij et al., 2011a], biodynamic feedthrough was measured with a
free control device, i.e., a stick that has at least one degree of freedom. In these studies the involuntary *position* deviations of the
control device were determined. When evaluated in the frequency
domain, the results obtained from the method which investigates
forces look very different from the results obtained for the method
which investigates positions. However, a relationship must obviously exist between these results, as they both describe the same
phenomenon, but from a different perspective. Both approaches
yield interesting and insightful results, and solving the apparent
ambiguity is therefore not simply a matter of selecting a 'preferred'
method. One should rather understand the relationship between
the two methods and their results.

As a first step to distinguish between these two different but related
dynamics, it was proposed in [Venrooij et al., 2013a] to label them

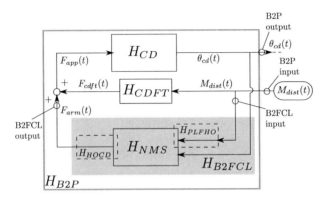

Figure 3.3: The dynamical elements that play a role in the response to motion disturbances: BDFT to positions (B2P) dynamics and BDFT to forces in closed-loop (B2FCL) dynamics.

BDFT to forces (B2F) and BDFT to positions (B2P) respectively. The latter type is elaborated here first.

Biodynamic feedthrough to positions

Biodynamic feedthrough to positions (B2P) is indicated in the outlined gray box in Fig. 3.3 and is expressed mathematically as:

$$H_{B2P}(s) = \frac{\theta_{cd}(s)}{M_{dist}(s)}$$

3.4

Note the similarity between the definition of H_{B2P} above and H_{FDFT} in Eq. 3.3, reflecting the fact that both FDFT and B2P describe the way a disturbance signal, a force disturbance and a motion disturbance respectively, results in control device deflections. The interpretation of B2P is therefore also similar to that of FDFT: it represents how external motion disturbances result in control device deflections (in [rad / (m/s^2)]). A large/high B2P means that an acceleration causes a large involuntary limb and control device deflection.

Biodynamic feedthrough to forces in closed-loop

A seemingly straightforward way of expressing biodynamic feed-through to forces (B2F) is:

$$H_{B2FCL}(s) = \frac{F_{arm}(s)}{M_{dist}(s)}$$

3.5

These transfer dynamics, indicated in the filled gray box in Fig. 3.3, are referred to as biodynamic feedthrough to forces *in closed-loop*, B2FCL. These dynamics describe the occurrence of forces due to BDFT effects in a closed-loop between NMS and CD. B2FCL (in [N / (m/s^2)]) is the force equivalent of B2P, by using the force signal $F_{arm}(t)$ as output instead of the position signal $\theta_{cd}(t)$. A large/high B2FCL means an acceleration causes a large involuntary force.

A complication with B2FCL is that its output, $F_{arm}(t)$, is enclosed in a closed-loop system and contains contributions from both the $M_{dist}(t)$ and the $\theta_{cd}(t)$ signals. As can be observed in Fig. 3.3, the H_{NMS} is modeled as a MISO system with two inputs, i.e., an $M_{dist}(t)$ 'channel' and a $\theta_{cd}(t)$ 'channel'. A possible way of dealing with this is by splitting the MISO block into two SISO blocks, as shown in Fig. 3.4, which gives rise to the concept of biodynamic feedthrough to forces *in open-loop*, B2FOL.

Biodynamic feedthrough to forces in open-loop

The two SISO NMS blocks, H_{NMS}^{mdist} and $H_{NMS}^{\theta cd}$, describe the dynamics of the neuromuscular system in response to motion disturbances and in response to control device deflections respectively. Note that the lower path in the B2FCL block in Fig 3.4 represents the inverse of the neuromuscular admittance H_{adm}^{-1}, just as in Fig. 3.2. The upper path is what can be labeled biodynamic feedthrough to forces in open-loop (B2FOL). From Fig. 3.4 it follows that:

$$H_{B2FOL}(s) = \frac{F_{b2fol}(s)}{M_{dist}(s)}$$

3.6

Just as B2FCL dynamics, the B2FOL dynamics describe the feed-through of accelerations to forces (in [N / (m/s^2)]). Again, a high

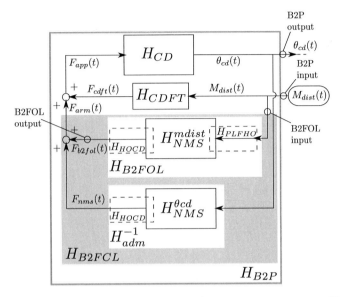

Figure 3.4: The B2FCL dynamics can be split into two SISO systems, H_{B2FOL} and H_{adm}^{-1}, describing the B2FOL dynamics and inverse of the admittance respectively.

B2FOL means an acceleration causes a large involuntary force. The difference between the OL and CL variant is which type of force, and thus which dynamics, is described. B2FCL is rather straightforward to obtain and hence an obvious type of BDFT dynamics to include. The B2FCL dynamics are related to the B2P dynamics through the known and static control device dynamics. This implies that B2FCL dynamics can be directly calculated from B2P dynamics and vice-versa (see Section 3.5). As a consequence the information contained in the B2FCL dynamics is not fundamentally different than the information contained in the B2P dynamics. As was mentioned already, the B2FCL dynamics have the complication of containing contributions from two inputs through a MISO system. This is not the case for the B2FOL dynamics, which makes it an insightful type of dynamics.

Where B2P yields a 'global picture' showing how accelerations end up as undesired control device positions, incorporating all the dynamics along the way, B2FOL shows the actual core of the BDFT problem: how accelerations directly lead to involuntary forces [Venrooij et al., 2013a]. Answering the question how B2FOL, B2FCL and B2P dynamics are related provides more insight in how BDFT works as a whole. It will be shown in Section 3.9 that B2FOL allows to make the BDFT dynamics independent from the control device dynamics, which has several benefits in, e.g., the modeling of BDFT dynamics. Another possible use of the B2FOL dynamics lies in BDFT mitigation: one way of mitigating BDFT is through model-based force cancellation [Sövényi and Gillespie, 2007]. This approach is based on inserting a canceling force or torque at the control device which counters the involuntary BDFT induced contribution of the force applied to the control device. The force that needs to be canceled here is the B2FOL force. That is, one needs to cancel the involuntary forces applied by the NMS as a results of $M_{dist}(t)$. More details regarding B2FOL dynamics and how it can be obtained are provided in Section 3.4.6.

The concept of B2F was already proposed in [Venrooij et al., 2013a], where a slightly different notation was used and no distinction was made between B2FCL and B2FOL. The more detailed and extended framework presented in this chapter would classify the type of biodynamic feedthrough dynamics investigated in [Venrooij et al., 2013a] as B2FOL dynamics.

Before moving on, it is important to note that the signal $F_{arm}(t)$ is now composed of two contributions, $F_{b2fol}(t)$ and $F_{nms}(t)$. The equation describing the admittance, Eq. 3.2, can now be refined to:

$$H_{adm}^{-1}(s) = \frac{F_{nms}(s)}{\theta_{cd}(s)} \qquad \text{3.7}$$

Control device feedthrough

The control device feedthrough (CDFT) forces are the inertial forces generated by the mass of the control device subjected to the PLF accelerations. Note that the CDFT dynamics are completely independent from the human operator dynamics. If the center of mass of the control device is approximately at the location where the HO applies control forces, these forces can be calculated through simple Newtonian dynamics:

$$F_{cdft}(s) = H_{CDFT}M_{dist}(s)$$
$$= m_{cd}M_{dist}(s)$$

3.8

and H_{CDFT} can be expressed as:

$$H_{CDFT}(s) = \frac{F_{cdft}(s)}{M_{dist}(s)} = m_{cd}$$

3.9

In this case, when the mass of the control device increases, so does the effect of CDFT. If the center of mass is not at the location where the HO applies control forces, the H_{CDFT} block may contain slightly different dynamics. CDFT can usually be obtained by simply subjecting a control device to acceleration disturbances without a human holding the device. The CDFT dynamics can then be determined from the transfer dynamics between the acceleration signal and the measured control device forces. It is important to note that CDFT only occurs for control devices with an offset between the control axis and the device's center of mass. In some control devices, e.g., steering wheels, the center of mass and the control axis are (roughly) aligned, which makes them insensitive to CDFT due to linear accelerations (other BDFT effects may still occur). For many other control devices this is not the case. Especially for long stick-type control devices with low stiffness, e.g., helicopter cyclic and collective, CDFT may play an important role.

As the effective mass of the control device m_{cd} may be relatively small, $F_{cdft}(t)$ may be insignificant compared to the limb's contribution $F_{arm}(t)$. Therefore it may be allowable to neglect the influence of CDFT in the BDFT analysis. In the following derivations CDFT will not be neglected. Section 3.6 will elaborate on the possible inaccuracies which are introduced if one does.

In many cases it is fairly straightforward to account for the effects of CDFT. Typically, $F_{app}(t)$ can be obtained through a force sensor. $F_{app}(t)$ includes contributions from $F_{arm}(t)$ and $F_{cdft}(t)$. By computing F_{cdft} through Eq. 3.8 one can obtain F_{arm}:

$$F_{arm}(t) = F_{app}(t) - F_{cdft}(t)$$

3.10

The above operation requires H_{CDFT} and $M_{dist}(t)$ to be accurately known, which can generally be assumed to be the case.

3.4   Obtaining BDFT dynamics from measurements

The different dynamical relationships introduced so far can be obtained from measurement. This section elaborates on how this can be done. This serves as an addition to the practical description provided in Chapter 2 of an experiment where B2P and admittance were measured.

3.4.1   The disturbance signals

To measure BDFT dynamics, a motion disturbance $M_{dist}(t)$ needs to be present. This signal should be strong enough to elicit involuntary control inputs in order to measure BDFT. The signal can be predefined by the experimenter (in open-loop experiments) or follow from a vehicle model that is under control by the subject (in closed-loop experiments).

In order to measure BDFT and neuromuscular admittance simultaneously, two disturbance signals are required: an acceleration disturbance $M_{dist}(t)$, applied to the PLF, and a force disturbance $F_{dist}(t)$ applied to the CD. As was shown in Chapter 2, using the acceleration disturbance $M_{dist}(t)$ the BDFT dynamics can be determined; force disturbance $F_{dist}(t)$ permits obtaining the neuromuscular admittance. In order to separate the contribution of the two disturbance signals in the analysis, the signals need to be separated in the frequency domain [Abbink, 2006; Venrooij et al., 2011a]. This requires the signals to be multi-sines with power at a limited number of frequencies. Separation in the frequency domain implies that, if the range of frequencies where the motion disturbance contains power is labeled ω_m and the range where the force disturbance is applied ω_f, there is no overlap between ω_m and ω_f.

Fig. 3.5 was taken from [Venrooij et al., 2013a] (and will be addressed in more detail in Chapter 4). The figure illustrates an example of disturbance signal design in a BDFT study. The figure shows the power-spectral-density (PSD) of $M_{dist}(j\omega_m)$ and $F_{dist}(j\omega_f)$ (with

j the imaginary unit). Both disturbance signals were multi-sines, defined in the frequency domain. In this case a range between 0.05 Hz and 21.5 Hz was selected for the force disturbance signal. For the motion disturbance signal a range between 0.1 and 21.5 Hz was selected. These frequency ranges allowed for capturing the relevant dynamics to study the BDFT dynamics in relation to the admittance. Different frequency ranges may also be used, depending on simulator and control device capabilities and experimental focus. For both disturbance signals, 62 frequency points were selected in the frequency range, without overlap between the two disturbance signals. To allow for frequency averaging, power was applied to two adjacent frequency points for each point [Schouten et al., 2008a], yielding 31 pairs of frequency points for each disturbance signal. In the analysis the results were calculated for each available frequency point, i.e., 62 frequencies, and then averaged for each pair, resulting in 31 data points. This procedure increases the reliability of the results [Schouten et al., 2008a].

A final note regarding the force disturbance signal and measuring admittance: the reduced power method [Mugge et al., 2007] should be used to construct the force disturbance signal $F_{dist}(t)$. In the example shown in Fig. 3.5, reduced power was used for frequencies of 0.7 Hz and higher. The details of the reduced power method are not discussed here, but in summary: it allows estimation of task-relevant neuromuscular dynamics over a wide frequency bandwidth, while avoiding the artifact of suppressed reflexive activity resulting from conventional perturbations over such a wide bandwidth. For more details, see [Mugge et al., 2007].

3.4.2 Neuromuscular admittance

In order to estimate the neuromuscular admittance from measurement data, the forces applied to the control device $F_{app}(t)$ and the control device positions $\theta_{cd}(t)$ should be measured. From $F_{app}(t)$ the force applied by the limb $F_{arm}(t)$ can be obtained (Eq. 3.10). It was already noted that $F_{arm}(t)$ is composed of two contributions, $F_{b2fol}(t)$ and $F_{nms}(t)$. The fact that these contributions cannot be directly measured does not pose a problem for the estimation of the

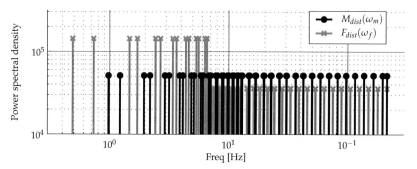

Figure 3.5: Power spectral density plot of disturbance signals $F_{dist}(t)$ and $M_{dist}(t)$.

admittance as long as the disturbance signal $F_{dist}(t)$ is separated in the frequency domain from the other disturbance signals or forcing functions. The procedure to estimate the admittance, which is presented below, includes a step where the Fourier transform of $F_{arm}(t)$ is multiplied with (the complex conjugate of) the Fourier transform of $F_{dist}(t)$. As the latter has non-zero values only for the frequencies ω_f, the multiplication only retains the data at the ω_f frequencies and removes everything else. This procedure allows us to use $F_{arm}(t)$ to calculate the neuromuscular admittance, without having to obtain $F_{nms}(t)$.

To calculate the neuromuscular admittance, $F_{arm}(t)$ and $\theta_{cd}(t)$ are transformed to the frequency domain using the fast Fourier transform (FFT):

$$\{F_{arm}(t), \theta_{cd}(t)\} \xrightarrow{FFT} \{F_{arm}(j\omega), \theta_{cd}(j\omega)\} \qquad \text{3.11}$$

As the neuromuscular system dynamics are embedded in a closed-loop configuration (see Fig. 3.2), a closed-loop identification technique is required to estimate the admittance [van der Helm et al., 2002]. In this closed loop $F_{arm}(t)$ and $\theta_{cd}(t)$ are dependent signals and the force disturbance signal $F_{dist}(t)$ is an independent signal. If $F_{dist}(t)$ is a multi-sine, it only has power on specific frequencies, i.e., on frequencies ω_f. Using the FFT of this signal the following cross-spectral densities can be estimated:

$$\hat{S}_{fdist,\theta}(j\omega_f) = F^*_{dist}(j\omega_f)\theta_{cd}(j\omega_f)$$ (3.12)

and

$$\hat{S}_{fdist,f}(j\omega_f) = F^*_{dist}(j\omega_f)F_{arm}(j\omega_f)$$ (3.13)

where the asterisk indicates the complex conjugate and the super-scripted hat symbolizes an estimate.

The estimate of the neuromuscular admittance, $\hat{H}_{adm}(j\omega_f)$, can now be calculated for frequencies ω_f as follows:

$$\hat{H}_{adm}(j\omega_f) = \frac{\hat{S}_{fdist,\theta}(j\omega_f)}{\hat{S}_{fdist,f}(j\omega_f)}$$ (3.14)

An important measure for the reliability of the estimate is the squared coherence $\hat{\Gamma}^2_{adm}$ [van der Helm et al., 2002]:

$$\hat{\Gamma}^2_{adm}(j\omega_f) = \frac{\left|\hat{S}_{fdist,\theta}(j\omega_f)\right|^2}{\hat{S}_{fdist,fdist}(j\omega_f)\hat{S}_{\theta,\theta}(j\omega_f)}$$ (3.15)

where $\hat{S}_{fdist,fdist}(j\omega_f)$ and $\hat{S}_{\theta,\theta}(j\omega_f)$ are the auto-spectral densities of $F_{dist}(t)$ and $\theta_{cd}(t)$ respectively:

$$\hat{S}_{fdist,fdist}(j\omega_f) = F^*_{dist}(j\omega_f)F_{dist}(j\omega_f)$$ (3.16)

$$\hat{S}_{\theta,\theta}(j\omega_f) = \theta^*_{cd}(j\omega_f)\theta_{cd}(j\omega_f)$$ (3.17)

The squared coherence is a measure for the signal-to-noise ratio (SNR) and thus for the linearity of the dynamic process. This function equals one when no non-linearities and no time-varying behavior are present. The function equals zero when no linear behavior is present at all.

3.4.3   Force disturbance feedthrough

The force disturbance feedthrough (FDFT) can be obtained from measurements in a similar way as the admittance. In this case the input is the force disturbance signal $F_{dist}(t)$ and the output is the control device deflection signal $\theta_{cd}(t)$ (see Fig. 3.2). So, the FDFT estimate can be calculated, at frequencies ω_f, as follows:

$$\hat{H}_{FDFT}(j\omega_f) = \frac{\hat{S}_{fdist,\theta}(j\omega_f)}{\hat{S}_{fdist,fdist}(j\omega_f)} \qquad \text{3.18}$$

The squared coherence for the FDFT:

$$\hat{\Gamma}^2_{FDFT}(j\omega_f) = \frac{\left|\hat{S}_{fdist,\theta}(j\omega_f)\right|^2}{\hat{S}_{fdist,fdist}(j\omega_f)\hat{S}_{\theta,\theta}(j\omega_f)} \qquad \text{3.19}$$

Note that the squared coherence for the FDFT is equal to the one for the admittance (Eq. 3.15).

3.4.4 Biodynamic feedthrough to positions

The biodynamic feedthrough to positions (B2P) is calculated in a manner analogous to the way the FDFT is obtained. As can be seen in Fig. 3.3, the B2P has as input the motion disturbance signal $M_{dist}(t)$ and as output the control device deflections $\theta_{cd}(t)$. Compare this with the FDFT, which has the force disturbance signal $F_{dist}(t)$ as input. By replacing $F_{dist}(t)$ with $M_{dist}(t)$ in Eqs. 3.18 and 3.19, we obtain:

$$\hat{H}_{B2P}(j\omega_m) = \frac{\hat{S}_{mdist,\theta}(j\omega_m)}{\hat{S}_{mdist,mdist}(j\omega_m)} \qquad \text{3.20}$$

and

$$\hat{\Gamma}^2_{B2P}(j\omega_m) = \frac{\left|\hat{S}_{mdist,\theta}(j\omega_m)\right|^2}{\hat{S}_{mdist,mdist}(j\omega_m)\hat{S}_{\theta,\theta}(j\omega_m)} \qquad \text{3.21}$$

Note that the B2P dynamics can only be determined for frequency range ω_m, where $M_{dist}(t)$ has power.

3.4.5 Biodynamic feedthrough to forces in closed-loop

By calculating the biodynamic feedthrough to forces in closed-loop (B2FCL), one obtains the dynamics between $M_{dist}(t)$ as input and

$F_{arm}(t)$ as output (see Fig. 3.3). These dynamics can be estimated as follows:

$$\hat{H}_{B2FCL}(j\omega_m) = \frac{\hat{S}_{mdist,f}(j\omega_m)}{\hat{S}_{mdist,mdist}(j\omega_m)}$$

3.22

and the squared coherence for this case:

$$\hat{\Gamma}^2_{B2FCL}(j\omega_m) = \frac{\left|\hat{S}_{mdist,f}(j\omega_m)\right|^2}{\hat{S}_{mdist,mdist}(j\omega_m)\hat{S}_{f,f}(j\omega_m)}$$

3.23

3.4.6   Biodynamic feedthrough to forces in open-loop

As can be observed in Fig. 3.4, the biodynamic feedthrough to forces in open-loop (B2FOL) describe the dynamics between $M_{dist}(t)$ as input and $F_{b2fol}(t)$ as output. When attempting to obtain these dynamics from measurements one is faced with a problem: how to determine the force $F_{b2fol}(t)$? The force applied to the control device $F_{app}(t)$ can be measured. When $H_{CDFT}(s)$ is known, $F_{arm}(t)$ can be determined through Eq. 3.10. However, $F_{arm}(t)$ is still the sum of $F_{nms}(t)$ and $F_{b2fol}(t)$. As both originate from the human neuromuscular system there is no direct way of separating these two contributions. If one would like to determine the B2FOL, there are three known ways around this problem: (i) using (an estimate of) the FDFT dynamics, (ii) using (an estimate of) the admittance and (iii) fixing the control device in its position. All three options are elaborated below.

Calculating B2FOL using FDFT dynamics

Fig. 3.4 can be redrawn to obtain Fig. 3.6. No other changes were made than rearranging the blocks H_{NMS}^{mdist} and H_{CDFT} to the left and adding the influence of a force disturbance $F_{dist}(t)$. Note that the H_{CDFT} and H_{CD} blocks are part of the control device and the $H_{NMS}^{\theta cd}$ and H_{NMS}^{mdist} blocks are part of the human operator. It was already shown how the FDFT dynamics can be obtained on ω_f (Eq. 3.18). Recall that the FDFT dynamics describe the transfer dynamics of force disturbances through the combined dynamics of the NMS and

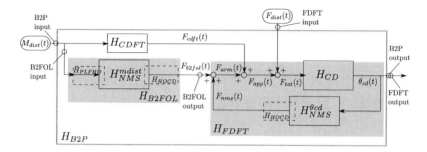

Figure 3.6: The B2P dynamics are a combination of B2FOL, CDFT and FDFT dynamics. Compare this figure with Fig. 3.4.

CD, resulting in limb/CD deflections (Fig. 3.2). From Eq. 3.3 we obtain:

$$\theta_{cd}(j\omega_f) = H_{FDFT}(j\omega_f)F_{dist}(j\omega_f) \qquad \text{3.24}$$

When, in addition to the force disturbance $F_{dist}(t)$, also a motion disturbance $M_{dist}(t)$ is applied, $F_{dist}(t)$ is no longer the only force disturbance present. In Fig. 3.6 it can be observed that $M_{dist}(t)$ causes two additional forces to work on the combined system of the HO and CD: $F_{b2fol}(t)$ and $F_{cdft}(t)$. It is important to note that, although the source of these latter forces is different, they can still be regarded as force disturbance and cause CD deflections in the same way as $F_{dist}(t)$ does. Hence, we can write:

$$\theta_{cd}(j\omega_m) = H_{FDFT}(j\omega_m)\left(F_{b2fol}(j\omega_m) + F_{cdft}(j\omega_m)\right) \qquad \text{3.25}$$

The equation above holds for frequencies ω_m, where $M_{dist}(t)$ has power. It also holds that (dropping $j\omega_m$ for brevity):

$$\theta_{cd} = H_{FDFT}\left(H_{B2FOL} + H_{CDFT}\right)M_{dist} \qquad \text{3.26}$$

Recalling from Eq. 3.4 that:

$$\theta_{cd} = H_{B2P}M_{dist} \qquad \text{3.27}$$

we obtain:

$$H_{B2P} = H_{FDFT}\left(H_{B2FOL} + H_{CDFT}\right) \qquad \text{3.28}$$

and an expression for H_{B2FOL} follows:

$$H_{B2FOL}(j\omega_m) = \frac{H_{B2P}(j\omega_m) - H_{FDFT}(j\omega_m)H_{CDFT}(j\omega_m)}{H_{FDFT}(j\omega_m)} \quad \boxed{3.29}$$

Note that in order to calculate $H_{B2FOL}(j\omega_m)$ the FDFT dynamics need to be known on frequency range ω_m. However, H_{FDFT} is measured using a force disturbance $F_{dist}(t)$ with frequency range ω_f, providing $H_{FDFT}(j\omega_f)$. A possible way to obtain $H_{FDFT}(j\omega_m)$ is to interpolate $H_{FDFT}(j\omega_f)$ to the frequency range ω_m. This approach was taken in [Venrooij et al., 2013a]. Note that this can only be done if ω_m is contained within ω_f. Possibly a more accurate method would be to create a model describing H_{FDFT} on ω_f and use this model to obtain $H_{FDFT}(j\omega_m)$ by evaluating the dynamics on frequencies ω_m.

Calculating B2FOL using admittance

From Eq. 3.5 and Fig. 3.6 it follows that (dropping $j\omega_m$):

$$\begin{aligned} H_{B2FCL} &= \frac{F_{arm}}{M_{dist}} \\ &= \frac{F_{b2fol} + F_{nms}}{M_{dist}} \\ &= H_{B2FOL} + \frac{F_{nms}}{M_{dist}} \end{aligned} \quad \boxed{3.30}$$

which can also be written as:

$$\begin{aligned} H_{B2FCL} &= H_{B2FOL} + \frac{F_{nms}}{\theta_{cd}}\frac{\theta_{cd}}{M_{dist}} \\ &= H_{B2FOL} + \frac{H_{B2P}}{H_{adm}} \end{aligned} \quad \boxed{3.31}$$

from which follows an alternative way of calculating B2FOL, now using an estimate of the B2P, B2FCL and admittance:

$$H_{B2FOL}(j\omega_m) = H_{B2FCL}(j\omega_m) - \frac{H_{B2P}(j\omega_m)}{H_{adm}(j\omega_m)} \quad \boxed{3.32}$$

It can be shown that Eq. 3.32 is equivalent to Eq. 3.29. The difference between this and the previous way of calculating B2FOL dynamics resides in the dynamics that are employed for the calculation, but results of both methods are identical. Eq. 3.29 was already used in a previous study with success [Venrooij et al., 2013a]. The relationship shown in Eq. 3.32 is novel and emerging from the framework presented here. It should be noted that an estimate of the admittance on ω_m is required, equivalent to the estimate of FDFT required on ω_m in the previous method.

Measuring B2FOL by fixing the stick

The third method of obtaining B2FOL is arguably more direct. By fixing the CD such that it cannot move, B2FOL can be measured directly. In this case, $\theta_{cd}(t)$ is always zero, and thus so is $F_{nms}(t)$, see Fig. 3.6. Hence, in this case:

$$F_{app}(t) = F_{b2fol}(t) + F_{cdft}(t) \qquad \text{3.33}$$

As $F_{cdft}(t)$ can be easily determined, direct access is obtained to the signal $F_{b2fol}(t)$. Note that in the case of a fixed control device H_{B2FCL} and H_{B2FOL} are equal [Venrooij et al., 2013a] and therefore B2FOL can be calculated using Eq. 3.22. The approach of fixing the control device was used successfully to obtain B2FOL models for model-based BDFT cancellation in [Sövényi and Gillespie, 2007]. However, in many practical cases, fixing the control device may not be a viable option. In those cases, the indirect methods using the FDFT dynamics (Eq. 3.29) or admittance (Eq. 3.32) may be preferred.

3.5 BDFT relationships

The different types of BDFT dynamics introduced in the previous sections are interrelated. The following presents the derivations which expose these relationships.

3.5.1 Force disturbance feedthrough relationships

From Fig. 3.2 the relationship between FDFT, CD and admittance can be derived (dropping the s for brevity):

$$\theta_{cd} = H_{FDFT}F_{dist}$$
$$\theta_{cd} = H_{CD}\left(F_{dist} + F_{arm}\right)$$
$$\theta_{cd} = H_{CD}\left(F_{dist} + H_{adm}^{-1}\theta_{cd}\right) \qquad (3.34)$$
$$\theta_{cd} = \frac{H_{CD}}{1 - H_{CD}H_{adm}^{-1}}F_{dist}$$

from which follows the interrelation between CD, admittance and FDFT dynamics:

$$H_{FDFT} = \frac{H_{CD}}{1 - H_{CD}H_{adm}^{-1}} = \frac{H_{adm}H_{CD}}{H_{adm} - H_{CD}} \qquad (3.35)$$

FDFT dynamics are thus a combination of the admittance and control device dynamics. From Eq. 3.35 it follows which of these dynamics governs the FDFT dynamics. If for example, $|H_{CD}(s)| \ll |H_{adm}(s)|$, implying that the control device has much stiffer dynamics than the human arm, Eq. 3.35 reduces to:

$$H_{FDFT} \approx \frac{H_{adm}H_{CD}}{H_{adm}} = H_{CD} \qquad (3.36)$$

The relationship was already empirically observed in [Venrooij et al., 2013a] and will be revisited in Section 3.8.

3.5.2 Relationship between B2P and B2FCL

Using Fig. 3.3 and Eqs. 3.4 and 3.5 the relationship between B2P and B2FCL can be derived (again, dropping the s):

$$H_{B2P}M_{dist} = \theta_{cd}$$
$$= H_{CD}\left(F_{cdft} + F_{arm}\right) \qquad (3.37)$$
$$= H_{CD}\left(H_{CDFT}M_{dist} + H_{B2FCL}M_{dist}\right)$$

so

$$H_{B2P}(s) = H_{CD}(s)\,(H_{CDFT}(s) + H_{B2FCL}(s)) \qquad \boxed{3.38}$$

The above indicates that B2P dynamics relates to B2FCL dynamics through static (and known) CD and CDFT dynamics. Hence, when the B2FCL is known, B2P can easily be calculated and vice-versa.

3.5.3 Relationship between B2P and B2FOL

A relationship between B2P and B2FOL was already obtained in Eq. 3.28. Here a more formal derivation is presented using Fig. 3.4 and Eq. 3.37:

$$
\begin{aligned}
H_{B2P}M_{dist} &= H_{CD}(F_{cdft} + F_{arm}) \\
&= H_{CD}(H_{CDFT}M_{dist} + F_{b2fol} + F_{arm}) \\
&= H_{CD}(H_{CDFT}M_{dist} + H_{B2FOL}M_{dist} \\
&\quad + H_{adm}^{-1}\theta_{cd}) \\
&= H_{CD}(H_{CDFT}M_{dist} + H_{B2FOL}M_{dist} \\
&\quad + H_{adm}^{-1}H_{B2P}M_{dist})
\end{aligned}
\qquad \boxed{3.39}
$$

from which follows:

$$
\begin{aligned}
H_{B2P} &= \frac{H_{CD}}{1 - H_{CD}H_{adm}^{-1}}(H_{CDFT} + H_{B2FOL}) \\
&= \frac{H_{adm}H_{CD}}{H_{adm} - H_{CD}}(H_{CDFT} + H_{B2FOL})
\end{aligned}
\qquad \boxed{3.40}
$$

so, by using Eq. 3.35:

$$H_{B2P}(s) = H_{FDFT}(s)\,(H_{CDFT}(s) + H_{B2FOL}(s)) \qquad \boxed{3.41}$$

Note that this result is equal to Eq. 3.28.

3.5.4 Relationship between B2FCL and B2FOL

The relationship between the two types of B2F, i.e., B2FCL and B2FOL, can be obtained through various ways. For example, by equating Eq. 3.38 and Eq. 3.41. Here, the more formal derivation is shown, based on Fig. 3.4 and Eq. 3.5:

$$
\begin{aligned}
H_{B2FCL}M_{dist} =& F_{arm} \\
=& F_{b2fol} + F_{nms} \\
=& H_{B2FOL}M_{dist} + H_{adm}^{-1}H_{CD}F_{app} \\
=& H_{B2FOL}M_{dist} + H_{adm}^{-1}H_{CD}F_{cdft} \\
& + H_{adm}^{-1}H_{CD}F_{arm} \\
=& H_{B2FOL}M_{dist} \\
& + H_{adm}^{-1}H_{CD}H_{CDFT}M_{dist} \\
& + H_{adm}^{-1}H_{CD}H_{B2FCL}M_{dist}
\end{aligned}
$$

(3.42)

from which follows:

$$
\begin{aligned}
H_{B2FCL}(1 - H_{CD}H_{adm}^{-1}) =& H_{B2FOL} \\
& + H_{adm}^{-1}H_{CD}H_{CDFT}
\end{aligned}
$$

(3.43)

so:

$$
\begin{aligned}
H_{B2FCL}(s) =& \frac{H_{adm}(s)}{H_{adm}(s) - H_{CD}(s)}H_{B2FOL}(s) \\
& + \frac{H_{CD}(s)}{H_{adm}(s) - H_{CD}}H_{CDFT}(s)
\end{aligned}
$$

(3.44)

This implies that the two forms of B2F are related through admittance, which is variable, and the static CD and CDFT dynamics.

The implications of this and other relationships presented here are not only theoretical. They provide us with an important insight into how one type of dynamics relates to others, which allows for, amongst other things, calculating one from others. An example of how these relationships can be applied in practice is shown in Section 3.9, where BDFT dynamics measured with one setting of CD dynamics is converted to BDFT dynamics obtained for different CD dynamics.

3.6   Neglecting control device feedthrough

As the mass and/or eccentricity of the center of mass may be small in some control devices, the influence of control device feedthrough (CDFT) may be neglected. In those cases, most of the relationships proposed in the previous sections can be simplified. When it is assumed that:

$$F_{cdft} = 0 \tag{3.45}$$

it follows that:

$$F_{app}(t) = F_{arm}(t) \tag{3.46}$$

meaning that the force applied by the HO becomes directly measurable. This simplifies the way B2FOL can be obtained, as in this case Eq. 3.29 reduces to:

$$H_{B2FOL}^+(j\omega_m) = \frac{H_{B2P}(j\omega_m)}{H_{FDFT}(j\omega_m)} \tag{3.47}$$

In order to distinguish between dynamics where CDFT is not neglected and dynamics where CDFT is neglected the superscripted $^+$ is added to the latter.

When fixing the control device, Eq. 3.33 becomes:

$$F_{app}(t) = F_{b2fol}(t) \tag{3.48}$$

When neglecting control device feedthrough, the relationship between B2P and B2FCL (Eq. 3.38) becomes:

$$H_{B2P}(s) = H_{CD}(s)H_{B2FCL}^+(s) \tag{3.49}$$

and the relationship between B2P and B2FOL (Eq. 3.41) simplifies to:

$$H_{B2P}(s) = H_{FDFT}H_{B2FOL}^+(s) \tag{3.50}$$

Finally, the relationship between B2FCL and B2FOL (Eq. 3.44) becomes:

$$H_{B2FCL}^+(s) = \frac{H_{adm}}{H_{adm} - H_{CD}}H_{B2FOL}^+(s) \tag{3.51}$$

So what does neglecting control device feedthrough mean for the accuracy of the results? The answer to that question depends not only on the control device feedthrough dynamics themselves, but also, and maybe more importantly, on which dynamics are considered. To understand the exact influence of neglecting control device feedthrough, it is important to realize that *neglecting* CDFT actually means *lumping* any measured CDFT dynamics which are physically present in the system with other dynamics. By neglecting CDFT, the effects do not disappear from the measurements, they are merely attributed to other sources. For example, by calculating B2FOL using Eq. 3.47 instead of Eq. 3.29, any CDFT effects present will be included in the B2FOL dynamics. This means the B2FOL dynamics are, in fact, over-estimated. The same holds when calculating B2FCL dynamics through Eq. 3.22, without correcting $F_{arm}(t)$ for $F_{cdft}(t)$. Also here, all possible CDFT effects – small or large – will be lumped inside the B2FCL dynamics, leading to an over-estimation.

The effect of neglecting CDFT dynamics can be expressed mathematically as:

$$H^+_{B2FOL}(s) = H_{B2FOL}(s) + H_{CDFT} \qquad \text{3.52}$$

and

$$H^+_{B2FCL}(s) = H_{B2FCL}(s) + H_{CDFT} \qquad \text{3.53}$$

The superscripted $^+$ indicates that CDFT dynamics are neglected, leading to an over-estimation.

Figs. 3.3 and 3.4 provide a visual intuition on the effect of neglecting CDFT: the H_{CDFT} block is lumped with the H_{B2FCL} or H_{B2FOL} respectively. Hence, neglecting CDFT dynamics means in practice that one is incorrectly attributing CDFT dynamics, belonging to the CD, to the neuromuscular system of the HO. An alternative way of interpreting 'neglecting' CDFT dynamics is 'not correcting' the B2FOL or B2FCL for them. In the following, the B2FOL and B2FCL dynamics will be referred to as 'corrected' or 'uncorrected' if CDFT was, respectively, accounted for or neglected. Section 3.8.4 will address the influence of correcting or not correcting for the CDFT dynamics in more detail.

Finally, it is important to note that the B2P dynamics are not influenced by the choice whether to correct or not correct for CDFT, as long as this choice is made consistently. Calculating B2P dynamics using Eq. 3.49 or Eq. 3.50 is exact, as long as B2FCL and B2FOL were not corrected for CDFT dynamics either, because in that case all CDFT dynamics are simply lumped into those dynamics. Hence, when constructing a B2P model from B2FCL dynamics, as will be done in Chapter 5 and Chapter 8, use can be made of the uncorrected B2FCL dynamics without influencing the quality of the model.

3.7 Validating the framework

Some of the concepts introduced in context of the BDFT framework have been proposed, applied, validated or demonstrated in the literature. Neuromuscular admittance has been widely used in many different studies (e.g., [Abbink, 2006; Mugge et al., 2007; van der Helm et al., 2002]). Most of the other concepts are proposed and discussed in the publications related to this thesis, although often using slightly different nomenclature and often not in full detail. For example, the dependency of B2P on neuromuscular admittance was demonstrated in [Venrooij et al., 2011a]; in [Venrooij et al., 2011a] the term 'BDFT' was used for what is referred to here as 'B2P'. The concept of FDFT was introduced in [Venrooij et al., 2010a] and elaborated in [Venrooij et al., 2013a]. These papers furthermore proposed the concept of B2P and B2F dynamics and also here slightly different nomenclature was used: in [Venrooij et al., 2010a] B2FOL was labeled 'motion disturbance feedthrough (MDFT)'. In [Venrooij et al., 2013a] B2F was used, but no distinction was made between B2FCL and B2FOL. In the light of the more detailed and extended discussion here, the type proposed, described and investigated in [Venrooij et al., 2013a] was B2FOL. Except for the nomenclature, all findings of both papers still hold. [Venrooij et al., 2013a] validated the concept of B2FOL and its relationship to B2P (Eq. 3.47). The concept of B2FCL was first proposed in [Venrooij et al., 2014b]. However, no details were provided about its relationship to B2FOL. The relationship between B2P and B2FCL was shown and used to construct a B2P model from B2FCL data

(Eq. 3.49). In all the mentioned papers, control device feedthrough was neglected in the analysis.

The BDFT framework proposed here completes and adds to the previously mentioned studies. Hence, a full validation of every aspect is not required. Instead, the validations will be limited to the following:

- The relationship between FDFT dynamics and the admittance and control device dynamics (Eq. 3.35).

- The relationship between B2P and B2FCL dynamics (Eq. 3.38).

- The relationship between B2P and B2FOL dynamics (Eq. 3.41).

- The relationship between B2FCL and B2FOL dynamics (Eq. 3.44).

- The approach to calculate B2FOL dynamics using FDFT dynamics (Eq. 3.47).

The data used to validate the framework was obtained in a study described in detail in [Venrooij et al., 2013a], which will be addressed in Chapter 4. In this study twelve subjects participated. For each, the BDFT and neuromuscular admittance were measured simultaneously, using a side-stick as control device. No armrest was present during the measurements. The measurements were performed in lateral direction (side-to-side) for three different control tasks: a position task (PT) or 'stiff task', with the instruction to minimize the position of stick, i.e., 'resist forces', a force task (FT) or 'compliant task', with the instruction to minimize the force applied to the stick, i.e., 'yield to forces', and a relax task (RT), with the instruction to relax the arm, i.e., 'ignore forces'. It was shown in Chapter 2 that for these tasks, human operators vary both their neuromuscular admittance and B2P dynamics.

3.7.1   Validating the relationships

Fig. 3.7 shows a comparison between measured and calculated dynamics, for the three different control tasks (PT, RT, FT). The data shown were obtained by averaging over all subjects. The figure

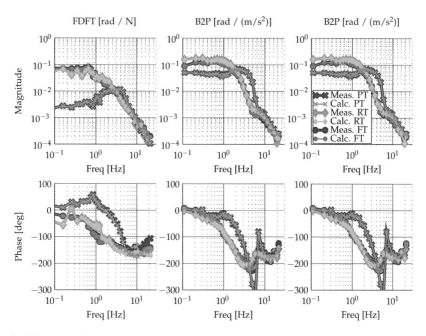

Figure 3.7: Validation of the proposed relationships. From left to right can be observed: a comparison between measured FDFT (Eq. 3.18) and calculated FDFT (Eq. 3.35), a comparison between measured B2P (Eq. 3.20) and calculated B2P through B2FCL (Eq. 3.38) and a comparison between measured B2P (Eq. 3.20) and calculated B2P through B2FOL (Eq. 3.41). The results show all relationships are accurate.

shows, from left to right, a comparison between measured FDFT (Eq. 3.18) and calculated FDFT (Eq. 3.35), a comparison between measured B2P (Eq. 3.20) and calculated B2P through B2FCL (Eq. 3.38) and a comparison between measured B2P (Eq. 3.20) and calculated B2P through B2FOL (Eq. 3.41). These figures cover the first three validation points listed above. As the measured and calculated dynamics are almost indistinguishable from each other, the results validate the proposed relationships.

Fig. 3.8 shows a comparison between measured B2FCL (Eq. 3.22)

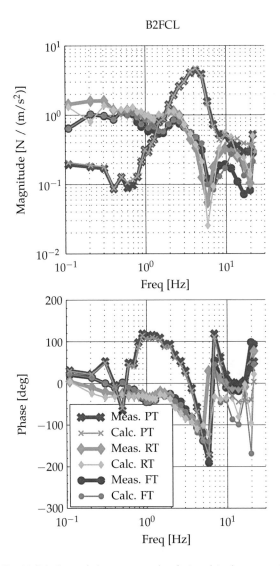

Figure 3.8: Validation of the proposed relationship between B2FCL and B2FOL. The figure shows a comparison between measured B2FCL (Eq. 3.22) and calculated B2FCL through B2FOL (Eq. 3.44). The results show the relationship is accurate, the differences are caused by the interpolation step required in the determination of B2FOL.

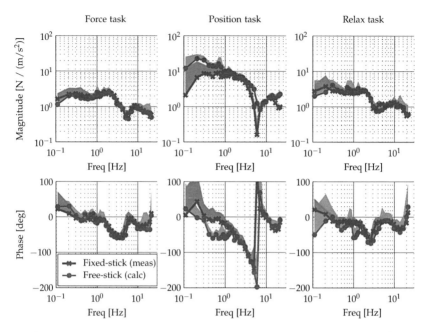

Figure 3.9: Comparison between measured and calculated B2FOL (H^+_{B2FOL}) dynamics. The lines indicate the mean; the colored bands indicate the (positive) standard deviation (mean + 1 SD).

and calculated B2FCL through B2FOL (Eq. 3.44). This plot addresses the fourth point of the listed validations. Some small differences can be observed between the measured and calculated B2FCL dynamics. These differences are caused by the interpolation step required to calculate B2FOL, i.e., $H_{FDFT}(j\omega_f)$ was interpolated to obtain $H_{FDFT}(j\omega_m)$, which introduces inaccuracies.

3.7.2 Validating the approach to calculate B2FOL dynamics

It was proposed that B2FOL dynamics can be determined by calculation, from B2P and FDFT data, using either Eq. 3.29 or Eq. 3.47 (respectively correcting or not correcting for CDFT). This section aims to provide evidence that this approach is indeed valid by comparing *measured* B2FOL dynamics, obtained using a fixed control device, with *calculated* B2FOL dynamics.

Five subjects participated in a validation experiment, where two settings for the control device dynamics were used: in the *fixed-stick* condition, the control device was fixed in its upright position and the force was measured that subjects applied while being subjected to a motion disturbance, allowing direct measurement of B2FOL (as was done in [Sövényi and Gillespie, 2007]). In the *free-stick* condition, the stick was free to move in lateral direction, which allowed obtaining B2P and FDFT through measurement and B2FOL by calculation, using Eq. 3.47. By comparing the results of these two methods of obtaining B2FOL, the proposed method of relating B2P, B2FOL and FDFT can be validated.

Measurements were performed for the three classical tasks (PT, FT and RT). Note that in the fixed-stick condition no force disturbance signal was applied, and therefore the FDFT dynamics and admittance could not be measured. In this condition the subjects were instructed to perform 'stiff', 'compliant' and 'relaxed' behavior for the PT, FT and RT condition, respectively. However, the absence of a force disturbance and control device deflections made the execution of these tasks harder. It is likely that control device deflections actually help the operator in the task execution, for example, by enabling or enhancing the use of reflexes. The absence of these control device deflections might have resulted in suboptimal performance, especially for the position task, where reflexes (e.g., muscle spindles and Golgi tendon organs) are employed to obtain maximum stiffness [Abbink, 2006].

Fig. 3.9 shows the comparison of the experimental results. The red lines and lighter red bands show the mean and standard deviation (SD) of the results obtained from the free-stick condition, where B2FOL was calculated. The blue lines and lighter blue bands show the mean and standard deviation of the results obtained in the fixed-stick condition, where B2FOL was measured directly. The results shown here were obtained without correcting for CDFT dynamics (i.e., H_{B2FOL}^{+} is shown). Note that the *relative* differences, which are of interest here, would be identical after correcting for CDFT as the operation would change both dynamics in exactly the same way. Clearly, the calculated B2FOL dynamics are very similar to the measured B2FOL dynamics. For the force task and relax task, both the magnitude and phase results are almost identical. For the position task some differences can be observed, especially in the magnitude at the lower frequencies. Also around 4-6 Hz some differences are visible, but here both dynamics share the same general feature, i.e., a notch, only the notch steepness and minimum differ. As these differences only occur for the position task, it is hypothesized that they are due to differences in reflexive activity, caused by the absence of stick deflections in the fixed-stick condition, rather than to a systematic error in the way B2FOL was calculated. Hence, despite the minor differences, the results in Fig. 3.9 provide good evidence that the method proposed here to calculate B2FOL dynamics is valid.

3.8   Interpreting BDFT dynamics

So far, much effort has gone into proposing, deriving and validating relationships between different types of dynamics. This section aims to demonstrate what these dynamics actually represent and what the proposed relationships practically imply. Having a mathematically sound and experimentally validated framework is useful, but practical knowledge on how to interpret the 'raw' transfer dynamics data, obtained from an experiment, is invaluable. This section will present and discuss several experimentally obtained transfer functions, aiming to provide some intuition to what we can learn about BDFT by observing the different types of dynamics.

3.8.1   FDFT, admittance and control device dynamics

Fig. 3.10 shows the neuromuscular admittance, control device dynamics (determined in a dedicated measurement), FDFT dynamics and squared coherence for a typical subject measured in a static condition (i.e., without a motion disturbance present). The admittance and FDFT were calculated by averaging over the different repetitions (lines). The colored bands indicate the standard deviation. The high squared coherences (close to 1) indicate that the

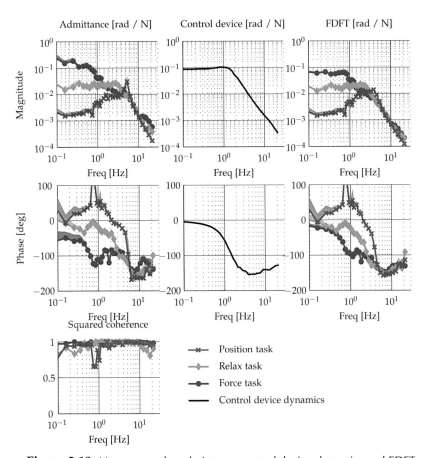

Figure 3.10: Neuromuscular admittance, control device dynamics and FDFT for one subject in static condition for the three control tasks. The lines indicate the mean over the different repetitions; the colored bands indicate the (positive) standard deviation (mean + 1 SD).

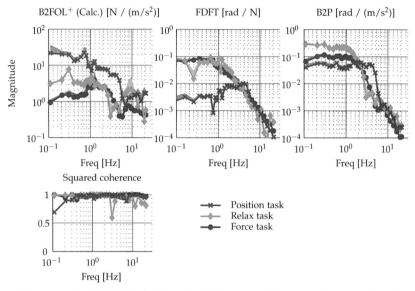

Figure 3.11: B2FOL$^+$, FDFT and B2P for one subject in motion condition for the three control tasks. The lines indicate the mean over the different repetitions; the colored bands indicate the (positive) standard deviation (mean + 1 SD).

results contain a high degree of linearity, which is an indication for the reliability of the measurement.

The admittance plot shows that the neuromuscular admittance is different in the three control tasks. As expected, the force task (FT) leads to a high admittance magnitude ('compliant' behavior), the position task (PT) to a low admittance magnitude ('stiff' behavior) and the relax task (RT) to an admittance level that lies in between. These results are similar to those found in previous studies into neuromuscular admittance of the arm [Lasschuit et al., 2008; Mugge et al., 2009].

Fig. 3.10 allows clarifying the interaction between admittance, control device dynamics and FDFT dynamics. The FDFT is a combination of the former two dynamics, as can be observed in Fig. 3.2 and Eq. 3.35. It was already mathematically shown that the FDFT dynamics is governed by the 'stiffest' dynamics of these two. This is similar to how two parallel mass-spring-damper systems would combine: if one would pull on one end of two mass-spring-damper systems arranged in parallel, one would primarily feel the influence of the 'stiffest' or 'heaviest' of the two systems. This superiority-of-stiffness principle becomes clear from the plots (where lower magnitude implies stiffer dynamics). Let's examine the dynamics for the force task first: up to approximately 1 Hz the control device dynamics are stiffer than the neuromuscular admittance. Hence, the control device dynamics govern the FDFT dynamics up to 1 Hz. It can be said that the FDFT 'flattens' in this region. Now, shifting our attention to the position task, the FDFT is governed by the neuromuscular admittance, as these dynamics are clearly stiffer than the control device dynamics. This holds across the full frequency range, except for the peak in the admittance occurring at 5 Hz. For this frequency, the admittance is high (compliant) and the FDFT is 'flattened' by the stiffer CD dynamics.

3.8.2   B2FOL, FDFT and B2P dynamics

Fig. 3.11 shows the B2FOL$^+$, FDFT and B2P dynamics for the same subject as in Fig. 3.10, but now for a motion condition, i.e., with a motion disturbance present. Note that the B2FOL dynamics were calculated from the FDFT and B2P estimates, through Eq. 3.47 (neglecting CDFT, i.e., H_{B2FOL}^+ is shown). Only magnitude information is shown in Fig. 3.11. Again, the lines represent the average over the different repetitions. The colored bands indicate the standard deviation. High squared coherences (shown for the B2P estimate) are observed, again an indication of a high degree of linearity in the results and reliability of the measurement.

When comparing the FDFT magnitude in Fig. 3.11 (motion condition) with that in Fig. 3.10 (static condition), it is observed that the results for the force task and position task in the two conditions are very similar. This suggests that the motion has no significant influence on the FDFT; this is in agreement with expectations, as the instructions for each task in the two conditions were identical. However, the FDFT magnitude during the relax task seems to have increased in the motion condition with respect to the static condition. This effect – the increase of the FDFT (and admittance) magnitude estimate in the relax task between the static and the motion condition – may be related to the effect observed and discussed in Chapter 2 and Chapter 4, i.e., the increase in admittance magnitude estimate for an increased force gain. For both observations, a possible and plausible cause for the increase in admittance is the influence of gravity on the admittance estimate. In the side-stick setup used here, gravity has the effect of magnifying deviations from the neutral point by pulling the arm down. The larger the deflection from the neutral point, the larger the role of the downward pointing gravitational force component. Such a magnification effect leads to a higher admittance estimate, even when the actual admittance of the arm remained the same. The problem encountered here with measuring RT dynamics in motion conditions is relevant and deserve further attention. However, as this thesis deals with the topic of BDFT, and not with estimating neuromuscular admittance, we propose a possible explanation in the form of the gravity hypothesis and leave the validation or rejection of this hypothesis as work for future studies.

Note that the B2P measurement shows a clear difference between all three tasks, confirming that the subject behaved differently across tasks. Similarly, one can observe that there are differences in B2FOL between each task as well. It was already shown in Chapter 2 that B2P dynamics are task dependent, here it is shown that B2FOL dynamics are task dependent as well.

Interestingly, the B2FOL is the highest for the position task, while this task leads to low admittance, low FDFT and low B2P. This can be explained as follows: B2FOL can be interpreted as an *acceleration-force coupling*, describing how accelerations result in forces applied to the control device. During a position task, the 'stiff' setting of the neuromuscular system realizes a tight coupling between the human body and control device in which accelerations, feeding through the human body, result in large forces on the control device, hence a high B2FOL. FDFT can be interpreted as the *force-position coupling*, describing how forces lead to deflections of the control device. A 'stiff' setting of the neuromuscular system implies that a force leads to a small position deviation, hence a low FDFT. B2P, seen in the right column of Fig. 3.11, is the combination of these two coupling effects and can be interpreted as the *acceleration-position coupling*. The force task, i.e., a compliant setting of the neuromuscular system, yields a low acceleration-force coupling (B2FOL) but a high force-position coupling (FDFT). So, in contrast to the position task, accelerations lead to smaller forces, but these smaller forces in turn lead to larger control device deflections.

3.8.3 The effects of changing the control device dynamics

In this section the following question is addressed: what happens to the BDFT dynamics when the control device dynamics are changed? An influence of control device dynamics on BDFT dynamics has been reported [McLeod and Griffin, 1989], but the exact effect can only be understood by observing the measurement results of B2P, B2FOL and FDFT dynamics. Just as in the previous sections, the B2FOL dynamics will be uncorrected for CDFT (i.e., H_{B2FOL}^{+} is shown).

Experimental data was obtained for twelve subjects using a control device (side-stick) with a stiffness of 0.2 N/deg. This stiffness setting will be referred to as the *compliant* setting. Another experiment was performed for three (different) subjects, now using a control device stiffness of 1.2 N/deg (the results of this experiment are only used for demonstration purposes, hence the small number of subjects). This stiffness setting will be referred to as the *stiff* setting.[a]

When control device dynamics are made stiffer it is to be expected that *B2P decreases* (as was also reported in McLeod and Griffin [1989]): with a stiffer control device the forces $F_{app}(t)$, including any involuntary BDFT component, simply results in a smaller position deviations $\theta_{cd}(t)$. Secondly, recall that the B2P dynamics are the product of the FDFT and uncorrected B2FOL^{+} dynamics (Eq. 3.50).

---

[a]Note that the stiff setting is different from the fixed setting used in Section 3.7.2 where the control device was fixed in its position. In fact, the fixed setting can be interpreted as the limit case of the stiff setting, where stiffness approaches infinity.

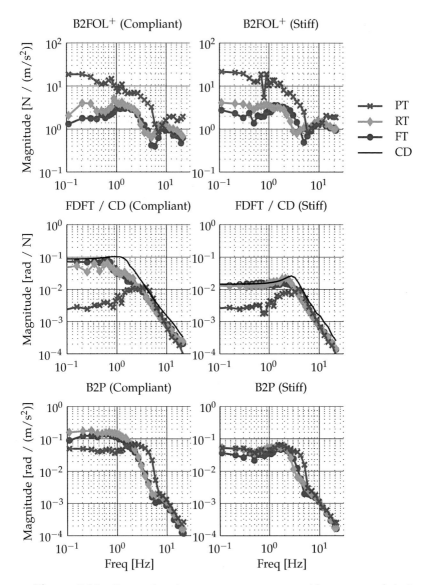

Figure 3.12: Comparison between measurements with two control device dynamics: a compliant control device (left column) and a stiff control device (right column).

In Fig. 3.6 it was shown that the control device dynamics are in-volved only in the FDFT dynamics and not in the B2FOL$^+$ dynam-ics. Hence, one would expect that *only the FDFT dynamics change* in response to a change in control device dynamics, while the B2FOL$^+$ dynamics remain unchanged. As the B2FOL$^+$ dynamics are the sum of B2FOL dynamics and CDFT dynamics (see Eq. 3.52), also the B2FOL dynamics remain unchanged. Thirdly, recall that the FDFT dynamics are the combination of the admittance and control device dynamics, where the stiffest of the two govern the FDFT dynamics. This 'flattening' effect was shown in Fig. 3.10. So, one expects that *a stiffer control device would 'flatten' the FDFT* more than a compliant control device as a larger part of the dynamics is gov-erned by the control device dynamics. Finally, note that such a change in FDFT will only occur if the control device dynamics are indeed stiffer than the neuromuscular admittance. This can be ex-pected to be the case the force task and relax task. If, on the other hand, the admittance is stiffer than the control device dynamics, as could be expected in the position task, the control device dynamics will not be of any influence on the FDFT dynamics. In that case, without a change in B2FOL$^{(+)}$ or a change in FDFT, *the B2P will remain the same for the position task.*
Concluding, based on the concepts of B2P, B2FOL, FDFT and their interactions, the following four hypotheses regarding the effect of a stiffer control device can be proposed:

- the effect will be visible in FDFT dynamics, not in the B2FOL$^{(+)}$ dynamics,

- the FDFT dynamics 'flatten' for the force and relax task,

- the FDFT and B2P decreases for the force and relax task, and

- the FDFT and B2P stays the same for the position task.

Fig. 3.12 shows the B2FOL$^+$, FDFT and B2P for the compliant and stiff control devices. Only the magnitude information is shown here, as it provides sufficient information to test our hypotheses.

The results confirm the hypotheses: first of all, there is no substantial difference between the B2FOL$^+$ for the two settings of the control device. Then, the FDFT is indeed 'flattened' for both the relax task and the force task. Clearly, for the lower frequencies the FDFT is governed by the control device dynamics (also shown), confirming again that the FDFT is governed by the stiffest element. Furthermore, one sees a reduction in the FDFT and B2P dynamics for the force and relax task, just as hypothesized. Finally, observe that the FDFT for the position task is indeed not affected by the change in control device stiffness. Also the B2P dynamics for the position task do not show a substantial difference, as both the B2FOL$^+$ and FDFT dynamics do not change due to the increased control device stiffness.

Before moving on, an additional note regarding these results is in place: although it is observed that stiffening the control device dynamics decrease the B2P dynamics, it should not be seen as a effective remedy for BDFT problems. A stiffer control device would require larger input forces for voluntary control. To maintain acceptable control forces the input gain of the stiffer control device needs to be increased, partially canceling the effects of the intended BDFT mitigation.

The foregoing discussion serves to provide an intuition on how the concepts of B2P, B2FOL and FDFT can aid in increasing the understanding of the BDFT process as a whole. Evaluating only B2P dynamics or only B2FOL dynamics, as was done in many previous studies, is often insufficient to understand the influence of a particular factor. By looking at B2P as a combination of B2FOL$^+$ and FDFT dynamics, however, one is able to understand and describe such influences better.

3.8.4   Neglecting CDFT dynamics

Fig. 3.13 shows the influence of neglecting CDFT on B2FCL dynamics. Recall that 'uncorrected' dynamics are those in which the influence of CDFT is neglected and 'corrected' dynamics are those where the CDFT was taken into account (i.e., removed). The

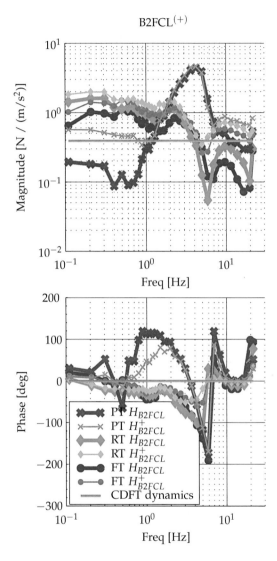

Figure 3.13: A comparison between B2FCL when correcting for control device feedthrough (H_{B2FCL}) and when neglecting control device feedthrough (H_{B2FCL}^{+}).

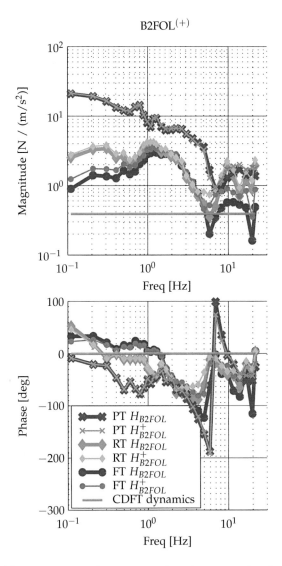

Figure 3.14: A comparison between B2FOL when correcting for control device feedthrough (H_{B2FOL}) and when neglecting control device feedthrough (H_{B2FOL}^+).

considerable differences observed between the corrected H_{B2FCL} and uncorrected H_{B2FCL}^+ are due to whether the CDFT dynamics is 'lumped' into the B2FCL dynamics or not. The difference between the corrected and uncorrected B2FCL dynamics is exactly the H_{CDFT} dynamics (see Eq. 3.53), i.e., the mass of the side-stick (0.39 kg), which are also indicated in the figure.

Fig. 3.14 shows a similar comparison, but now between corrected and uncorrected B2FOL dynamics. Visually, the differences appear to be much smaller, possibly negligible. However, it should be noted that the differences between H_{B2FOL} and H_{B2FOL}^+ are exactly the same as for the B2FCL dynamics (see Eq. 3.52), i.e., the mass of the CD, and only appear smaller because the larger magnitude of the B2FOL dynamics.

3.9   Applying framework knowledge: an example

In the previous section it was shown how 'raw' transfer dynamics can be interpreted. Amongst other things it was illustrated how a change in control device dynamics influences the different dynamics that play a role in BDFT. This section will use the same data, obtained for a compliant control device stiffness setting (0.2 N/deg) and a stiff control device stiffness setting (1.2 N/deg), to take on the following challenge: to *convert* the B2P dynamics measured with the compliant control device dynamics to the B2P dynamics obtained with stiff control device dynamics. This conversion can be accomplished by employing the mathematical relationships of the BDFT framework that have been proposed in this chapter. A successful conversion serves as an illustration of a practical application of the framework, and, equally important, it demonstrates that the influence of the CD dynamics on BDFT is now thoroughly understood.

The measured B2P dynamics for the two settings of the CD dynamics are plotted per task in Fig. 3.15 for comparison. Clearly, the B2P dynamics are very different for the two cases. In the following, the superscript ^{comp} refers to the data obtained with the compliant CD dynamics, ^{stiff} to those obtained with the stiff CD dynamics

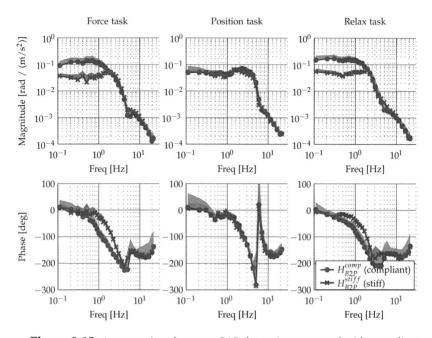

Figure 3.15: A comparison between B2P dynamics measured with compliant and stiff control device dynamics.

and ^{conv} to the converted dynamics. From measurements with the compliant CD, $H_{B2P}^{comp}(j\omega_m)$, $H_{CDFT}^{comp}(j\omega_m)$ and $H_{adm}^{comp}(j\omega_f)$ can be obtained. Using Eq. 3.29, H_{B2FOL}^{comp} can be calculated (which is corrected for CDFT dynamics). From the representation in Fig. 3.6 and the empirical results in Fig. 3.12, it follows that B2FOL dynamics are independent from CD dynamics, hence:

$$H_{B2FOL}^{conv}(j\omega_m) = H_{B2FOL}^{comp}(j\omega_m) \qquad \boxed{3.54}$$

And, as the mass of the control device and the eccentricity of the center of mass did not change by changing the stick's stiffness, also that:

$$H_{CDFT}^{conv}(j\omega_m) = H_{CDFT}^{comp}(j\omega_m) \qquad \boxed{3.55}$$

Furthermore, it is assumed that:

$$H_{adm}^{conv}(j\omega_m) = H_{adm}^{comp}(j\omega_m) \qquad \boxed{3.56}$$

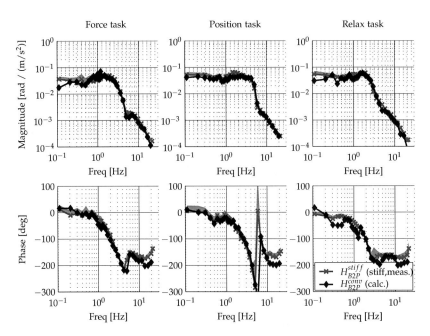

Figure 3.16: A comparison between measured B2P dynamics (with stiff CD dynamics) and calculated B2P dynamics (converted from data with compliant CD dynamics).

as the control device dynamics do not change the limb dynamics of the HO. Now, the converted FDFT dynamics $H_{FDFT}^{conv}(j\omega_f)$ can be obtained using Eq. 3.35:

$$H_{FDFT}^{conv}(j\omega_f) = \frac{H_{adm}^{conv}(j\omega_f) H_{CD}^{stiff}(j\omega_f)}{H_{adm}^{conv}(j\omega_f) - H_{CD}^{stiff}(j\omega_f)} \qquad \boxed{3.57}$$

which implies that the FDFT dynamics are recalculated using different control device dynamics. By interpolating $H_{FDFT}^{conv}(j\omega_f)$ on frequencies ω_m, $H_{FDFT}^{conv}(j\omega_m)$ is obtained.

Now, the converted B2P dynamics $H_{B2P}^{conv}(j\omega_m)$ can be calculated using Eq. 3.41:

$$H_{B2P}^{conv}(j\omega_m) = H_{FDFT}^{conv}(j\omega_m) \left(H_{CDFT}^{conv}(j\omega_m) + H_{B2FOL}^{conv}(j\omega_m) \right) \qquad \boxed{3.58}$$

The results are shown in Fig. 3.16, where it can be observed that the conversion was successful. The converted B2P dynamics are very close to the measured B2P dynamics. The minor differences can be attributed to two main causes: (i) the fact that the measurements were performed for two different subject groups (the 'comp' group consisting of 12 subjects, the 'stiff' group of only 3) (ii) the interpolation step required to obtain $H_{FDFT}^{conv}(j\omega_m)$ introduces inaccuracies (just as in Fig. 3.8). The latter issue can be largely solved by fitting a model on the FDFT data and evaluating the model on the required frequency range.

These results show that, using the proposed framework, the influence of CD dynamics on BDFT can be thoroughly understood and that, in addition, B2P dynamics measured with a particular CD dynamics can be converted to those for other CD dynamics.

3.10   Applying the framework to literature

One of the objectives of proposing a BDFT framework is to provide a common ground to study, discuss and understand BDFT and its related problems. To illustrate how this goal was reached three BDFT studies were selected from literature to be evaluated as case studies within the current framework.

3.10.1   Case study I: mitigating B2FOL dynamics

Sövényi and Gillespie [2007] proposed a model-based cancellation controller to mitigate the effect of BDFT. Their approach was to subject human operators to a lateral acceleration disturbance with the instruction to not give any cognitive input but 'simply hold the joystick handle'. During these measurements the control device was immobilized using a peg. In the analysis the transfer dynamics were obtained between acceleration signal $M_{dist}(t)$ and applied forces $F_{app}(t)$. Recall that by fixing the control device the applied force $F_{app}(t)$ is the sum of $F_{b2fol}(t)$ and $F_{cdft}(t)$ (see Eq. 3.33). Hence, the resulting transfer dynamics obtained in [Sövényi and Gillespie, 2007] described what we defined earlier as uncorrected B2FOL dynamics H_{B2FOL}^{+}. The paper made no distinction between B2FOL and B2P dynamics and referred to both as 'BDFT dynamics'. The task instruction used in [Sövényi and Gillespie, 2007] was very similar to the control task instruction of the relax task (RT) (see Section 3.7). The B2FOL dynamics obtained in [Sövényi and Gillespie, 2007] were approximated using an ARMA model. In a second part of the experiment, where the peg was removed allowing the control device to move, this model was used to 'reflect' the involuntary forces caused by BDFT. This approach can be represented in Fig. 3.6 by replacing $F_{dist}(t)$ with the *negative* output of the B2FOL model, which is an approximation of the sum of F_{b2fol} and F_{cdft}, i.e., the forces due to acceleration signal $M_{dist}(t)$.

The above illustrates how the work reported in [Sövényi and Gillespie, 2007] can be interpreted using the nomenclature of the proposed framework. In addition, in [Venrooij et al., 2010a] we showed that the B2FOL dynamics *measured* in [Sövényi and Gillespie, 2007] were similar to the B2FOL dynamics *calculated* in [Venrooij et al., 2010a], using Eq. 3.47. Despite the fact that in [Venrooij et al., 2010a] these dynamics were labeled 'motion disturbance feedthrough' or MDFT, a name which we would now suggest to no longer use, the comparison showed that the dynamics measured by Sövényi and Gillespie were indeed what we refer to as B2FOL dynamics obtained during a 'relax task' (RT).

3.10.2   Case study II: a vertical BDFT model

Mayo [1989] proposed a BDFT model for a helicopter pilot holding the collective lever, subjected to vertical accelerations. Two parameter sets were proposed, one obtained for ectomorphic (slim bone structure and muscle build) subjects and one obtained for mesomorphic (athletic bone structure and muscle build) subjects. The model aimed to account for differences between these somatotypes, i.e., for inter-subject variability. The variability that the model aimed to address are thus located in the NMS block of the BDFT system model (Fig. 3.1). The differences between somatotypes were expected to lead to differences in the dynamics of the neuromuscular dynamics. To describe the experimental results, the following transfer function pilot model was proposed, describing the absolute acceleration of the hand holding the collective as a function of the seat's vertical acceleration [Mayo, 1989]:

$$H_{mayo}(s) = \frac{b_1 s + b_2}{s^2 + a_1 s + a_2} \qquad \text{3.59}$$

where s represents the Laplace operator. The values of the four parameters a_1, a_2, b_1 and b_2 were found by fitting the transfer function on the data obtained for the two somatotypes. Note that the model described dynamics from vehicle accelerations (input) to hand accelerations (output). Similar transfer dynamics, from vehicle accelerations to the accelerations of body parts such as hand, shoulder, head and thorax, were reported in several other, mainly older, studies (e.g., [Allen et al., 1973] and [Jex and Magdaleno, 1978]). In the framework introduced here, such a dynamical relationship was not included. So far, only BDFT to forces (B2F) and BDFT to positions (B2P) were distinguished. In order to include the type of BDFT dynamics measured in [Mayo, 1989], one could decide to introduce BDFT to accelerations (B2A). Although this is not necessarily incorrect, it may be more insightful to establish a relationship between B2A and B2P. This was done in [Masarati et al., 2013] by exploiting the fact that acceleration (e.g., in $[m/s^2]$) is the second derivative of position (e.g., in [m]). Converting B2A to B2P can thus be done by adding two pseudo-integrators, $1/(s+c)$ (where parameter c can be used to eliminate drift and account for the pilot's ability to correct for low-frequency disturbances [Masarati et al., 2013]). These pseudo-integrators convert acceleration to position. In this case, the resulting position of the collective lever in [m] can be converted into a deflection in [rad], which is a more common unit for B2P measurements, by dividing by the length of the collective lever L. To obtain the relative deflection one needs to subtract the absolute acceleration of seat from the absolute acceleration of the hand, resulting in $H_{mayo}(s) - 1$. The relative deflection of the control device as a function of the acceleration can be thus written as (also see [Masarati et al., 2013] and [Venrooij et al., 2014b]):

$$H_{B2P}(s) = \frac{1}{s + c_1} \frac{1}{s + c_2} \frac{1}{L} \left(H_{mayo}(s) - 1 \right) \qquad \text{3.60}$$

The results in [Venrooij et al., 2014b] show that the conversion of Mayo's model provides accurate results, but within the frequency range the model was designed for and only for the RT. The same paper also extends and improves Mayo's model, illustrating how to obtain a B2P model from a B2A model. Such a conversion may be applied to other studies where B2A was measured, allowing for an appreciation of the results from previous BDFT research in a new context. Chapter 5 will show that a much more accurate model can be obtained by modeling B2FCL dynamics and converting the resulting model to a B2P model by multiplying with the control device dynamics (Eq. 3.38).

3.10.3   Case study III: BDFT in rotorcraft

A survey of pilot-structural coupling instabilities in naval rotorcraft is provided by Walden [2007]. The publication discusses modeling and mitigation techniques for Pilot Assisted Oscillations (PAOs). Also some case studies are discussed. Note that in [Walden, 2007] no mention is made of the term 'BDFT' or any of the other types of BDFT related dynamics proposed in the current chapter. Instead, the paper mentions the terms "limb-manipulator effect" or "limb-bobweight effect". Regarding mitigation techniques, a distinction is made between *flight restrictions*, applied "to prevent flight in known regimes where PAOs can occur" and *procedural mitigations*, which are "recommendations to modify the pilot's behavioral response to the PAO". Other mitigation techniques include *adaptations to the rotorcraft* itself, such as viscous stick dampers, rotor dampers and input notch filters. From the case studies it becomes clear that most PAOs were dealt with using notch filters (V-22B Osprey, AH-1Z Viper), adaptations of the control device(s) (CH-46E Sea Knight, V-22B Osprey), and through procedural mitigations, instructing the pilots to reduce the magnitude of the inputs (CH-46D/E Sea Knight, CH-53E Super Stallion) or release/diminish grip of a control device (SH-60B Black Hawk, V-22B Osprey).

In the framework proposed in the current study, these PAO examples and their mitigation techniques can all be incorporated in

Fig. 3.1. First of all, we are considering a closed-loop BDFT system here, where the CE and PLF represent the helicopter dynamics. Flight restrictions act as a mitigating mechanism simply by avoiding certain systems states $y_{cur}(t)$, which may cause PLF motions $M_{dist}(t)$ which trigger a PAO. Implementing notch filters means that $\theta_{cd}(t)$ is filtered before it enters the CE, removing BDFT induced inputs on a particular frequency. Note that also voluntary inputs at this frequency are removed. Furthermore, note that this mitigation technique does not target the occurrence of BDFT itself. It merely 'ignores' the input at frequencies that may excite a structural mode and cause a closed-loop oscillation. The adaptations made to the control device, such as adding extra mass or increasing damping and stiffness, change the H_{CD} and H_{CDFT} blocks in order to reduce BDFT in a similar way as was done in Section 3.9. Note that the effectiveness of this method may depend on the neuromuscular setting of the pilot. In case the pilot, under influence of stress or high workload, grips the control device firmly and exhibits a stiff neuromuscular setting (i.e., performs a PT), the influence of changing the control device dynamics may be limited or absent (see Fig. 3.15). Finally, the instruction to reduce grip on the control device effectively reduces the transfer of involuntary limb motion to involuntary applied forces (i.e., adapts the H_{NMS} and H_{HOCD} blocks). By doing so, the role of the HO in the control loop is effectively diminished, reducing both involuntary and voluntary control inputs. Finally, note that the proposed adaptation techniques cover all blocks of Fig. 3.1, except H_{PLFHO}. Possibly, there are undiscovered methods of reducing PAOs by altering the interface dynamics between the rotorcraft and the body of the pilot.

The discussion of the above studies is far from complete. They serve merely as brief illustrations of how the current framework can be applied. An in-depth re-interpretation of any study within the current framework would take up more space than is available here. There are many other BDFT studies that would serve equally well for comparison as the three that were selected. The above examples illustrate how methods and results of different studies can be interpreted using the framework proposed in the current chapter. The value of this is that it allows for a comparison across studies and a reinterpretation of results.

3.11 Conclusions

Despite the attention it has received from the scientific community over the past decades, biodynamic feedthrough is still recognized as a complex problem with many unanswered questions. Moreover, across the different studies there is little consensus on how to approach the problem in terms of definitions, nomenclature and mathematical descriptions. The fragmentation in BDFT literature obstructs comparing results obtained in existing studies and building on those studies to further our understanding of BDFT phenomena.

In this chapter a framework for BDFT analysis was proposed which aims to provide a tool in unifying what has been presented in the literature so far, but may also serve as a guideline for future BDFT research. In addition, the framework itself allows for gaining novel insights into BDFT phenomena.

It was presented how relevant signals can be obtained from measurement and how different BDFT dynamics can be derived from them. It was shown how BDFT can be dissected into several dynamical relationships and how each of them is relevant in understanding BDFT. In this chapter the following dynamics were introduced in detail:

- Biodynamic feedthrough to positions (B2P): the feedthrough of accelerations to involuntary control device deflections.

- Biodynamic feedthrough to forces (B2F): the feedthrough of accelerations to involuntary control forces, existing in both a closed-loop form (B2FCL) and an open-loop form (B2FOL).

- Control device feedthrough (CDFT): the transfer of accelerations through the control device mass.

It was shown how these dynamics relate to each other. Furthermore, it was shown how they are influenced by human body dynamics, which can be described by the neuromuscular admittance and force disturbance feedthrough (FDFT). The proposed relationships were validated using experimental results.

By observing the transfer function dynamics obtained in several experiments, an intuition was gained regarding the different dynamics and the implications of their relationships. It was shown that FDFT dynamics are governed by the stiffer of either the admittance or control device dynamics. It was shown that B2FOL can be interpreted as acceleration-force coupling, FDFT as force-position coupling and B2P as acceleration-position coupling. It was shown how the concepts of FDFT, B2P and B2FOL can help in understanding the influence of changes in the control device dynamics. Finally, the influence of neglecting CDFT dynamics on both B2FCL and B2FOL dynamic was demonstrated.

As an example of a practical application of the framework, it was demonstrated how the effect of control device dynamics on BDFT can be accurately predicted. It was shown how B2P dynamics can be transformed by recalculating the FDFT dynamics using control device dynamics of choice. The results showed that, using the proposed framework, the influence of CD dynamics on BDFT can be thoroughly understood and that, in addition, B2P dynamics measured with a particular control device dynamics can be converted to those for other control device dynamics. Finally, by applying the framework on three selected BDFT studies, it was shown how the approaches and results put forward in those studies can be reinterpreted into one general framework, leading to the following conclusions:

- The mitigation approach proposed by Sövényi and Gillespie [2007] was to measure uncorrected B2FOL – using the direct approach of fixing the control device's position – during a relax task. The approach of force reflection used in [Sövényi and Gillespie, 2007] can be represented directly in the BDFT system model that was presented in this chapter, showing that the approach can be interpreted using the nomenclature of the proposed framework.

- The BDFT model proposed by [Mayo, 1989] could be labeled as a BDFT to acceleration (B2A) model. Instead of including this type of BDFT dynamics in the framework, one can also convert the model to obtain a more practical B2P model, as was done in [Masarati et al., 2013].

- The survey of pilot-structural coupling instabilities in naval rotorcraft by Walden [2007] can be reformulated using the framework's nomenclature. It was shown that all the BDFT mitigation techniques mentioned in [Walden, 2007] can be represented in the BDFT system model. In addition, this analysis brings to light that one particular option for BDFT mitigation, i.e., adaptation of the PLF-HO interface, is not exploited in practice yet.

The framework suggested here is not final nor complete. As knowledge on BDFT increases, so may the limitations of the current framework become apparent. The framework should be considered open for future additions and extensions, paving the way to unify the efforts of individual research groups investigating BDFT related phenomena. Examples of effects that are currently not (yet) represented in the framework are the influence of cognitive control actions and the influence of an armrest. Additions to the framework on how to define, measure and describe these influences, would not only make the framework more complete but also increase our understanding. The current chapter aims to show that defining such a framework, already with the current level of understanding, is valuable, maybe even necessary to further the progress of BDFT research and, eventually, mitigate its detrimental effects.

Part II

Modeling
biodynamic feedthrough

CHAPTER 4

A physical biodynamic feedthrough model

This chapter presents a physical biodynamic feedthrough model based on neuromuscular principles. The model structure uses an established admittance model, describing limb dynamics, which is expanded to include control device dynamics and account for BDFT effects. The model serves primarily the purpose of increasing our understanding of the relationship between admittance and biodynamic feedthrough. The quality of the physical BDFT model was evaluated in the frequency and time domain. The results show that the physical BDFT model accurately describes BDFT dynamics obtained for different subjects and under different conditions, incorporating both between-subject and within-subject variability.

The contents of this chapter are based on:

Paper title   A Biodynamic Feedthrough Model Based on Neuro-
                   muscular Principles

Authors   Joost Venrooij, David A. Abbink, Mark Mulder, Mar-
                   inus M. van Paassen, Max Mulder, Frans C. T. van
                   der Helm, and Heinrich H. Bülthoff

Published in   IEEE Transactions on Cybernetics, Vol. 44, No. 7,
                   July 2014

4.1 Introduction

Among all the influencing factors in biodynamic feedthrough (BDFT) dynamics the human body dynamics are the most complex and possibly the most influential. The human body dynamics are of importance as it is through these dynamics that vehicle accelerations are transferred into involuntary control inputs. Studies showed that BDFT dynamics vary due to both static anthropometric features – such as body size and weight [Mayo, 1989] – and dynamic neuromuscular settings [Venrooij et al., 2011a]. It is known that humans can adapt their limb dynamics over time, through muscle co-contraction and modulation of reflexive activity, depending on factors such as task instruction, workload and fatigue [Abbink and Mulder, 2010; Mulder et al., 2011]. This renders BDFT a variable, dynamical relationship, not only varying between different persons (between-subject variability), but also within one person over time (within-subject variability).

A good measure to account for both between- and within-subject variability in limb dynamics has proven to be the neuromuscular admittance (or simply admittance). Admittance is a dynamic property of a limb characterized by the relationship between force input and position output [Abbink, 2006]. Accurate models of the admittance have been developed [Abbink, 2006; Mugge et al., 2009; Schouten et al., 2008b].

Also the modeling of biodynamic feedthrough has been studied for several decades. In many BDFT studies, the between-subject variability has been recognized (e.g., [Mayo, 1989; McLeod and Griffin, 1989; Sövényi and Gillespie, 2007]), but the issue of within-subject variability has largely been ignored. Many of the earlier BDFT studies were of an empirical nature and aimed at measuring the transfer dynamics between vehicle accelerations and limb motion (see [McLeod and Griffin, 1989] for a review). In the late 1960s and early 1970s, the first biodynamic feedthrough models appeared, e.g., [Jex and Magdaleno, 1978]. These models were primarily physical, parameterized models, based on empirical BDFT data. The main purpose of the development of these models was to increase the fundamental understanding of biodynamic feedthrough, by using a-priori knowledge and physical principles. This approach resulted in several models with a clear physical interpretation (e.g., [Gillespie et al., 1999; Hess, 1998; Jex and Magdaleno, 1978; Sirouspour and Salcudean, 2003]).

An important trait of BDFT modeling was discovered shortly after: its potential to be employed in biodynamic feedthrough mitigation. This insight started the development of 'model-based BDFT cancellation', which relies on a BDFT model to predict and correct for involuntary control input (see also Chapter 6). The models used for this purpose did no longer aim to reproduce actual physical phenomena, which resulted in the development of several black box BDFT models (e.g., [Mayo, 1989; Sövényi, 2005; Velger et al., 1984]). Black box models are characterized by a different modeling approach geared towards describing BDFT dynamics at 'end-point level', without aiming to include the underlying physical principles. A comprehensive discussion on physical and black box BDFT models is provided in Section 2.3 of Ref. [Sövényi, 2005] (p. 15). The main advantage provided by physical BDFT models over black box BDFT models is the additional insight gained in the physical processes underlying the BDFT phenomenon; however, this insight usually comes at a price. Since such models aim to describe the complexities of reality, they are usually more elaborate than their black box counterparts, which 'end-point level' description allows for lumping underlying physical principles together into efficient dynamical descriptions with only a few parameters. The parameter estimation of an elaborate physical model is often very challenging. Over-parameterization is a genuine risk with such complex models, which can make reliable parameter estimation impossible. On the other hand, black box models are often less versatile and flexible in comparison.

This chapter presents a novel physical BDFT modeling approach. The resulting physical BDFT model serves primarily the purpose of increasing the understanding of the relationship between admittance and biodynamic feedthrough. The challenging process of parameter estimation – argued to be the major drawback of a physical model – is handled by using a two-stage parameter estimation approach. First, the parameters of a well-known admittance model are estimated, using established techniques. Then, these parameter values are used within a BDFT model that was constructed by extending the admittance model, leaving only a limited number of parameters still to be estimated. It will be shown that the resulting physical BDFT model accurately describes BDFT dynamics obtained for different subjects and under different conditions, in both the frequency and time domain.

The chapter is structured as follows: Section 4.2 introduces the biodynamic feedthrough system model, containing all high-level elements of a general BDFT system, and discusses all other models used in this chapter. In Section 4.3 the models' transfer functions are presented. More modeling considerations are addressed in Section 4.4. The experiment is discussed in Section 4.5. The model's parameter estimation and time-domain validation is described in Sections 4.6 and 4.7. The chapter ends with results, discussion and conclusions in Sections 4.8–4.10.

4.2   The biodynamic feedthrough system model

The starting point for the development of the *physical* BDFT model is the BDFT *system* model, shown in Fig. 4.1, containing all the high-level elements in a general BDFT system. Not all elements of the BDFT system model will be included in the physical BDFT model developed in the current chapter. The input and output of the physical BDFT model developed here are indicated in the figure. An elaborate description of the BDFT system model was already provided in Chapter 3, which will not be repeated here.

The following points, discussed in Chapter 3 in detail, are of particular importance for the remainder of the current chapter:

- The neuromuscular admittance describes limb dynamics. Recall that in the BDFT system model the combination of the H_{NMS} block and the H_{HOCD} block forms the *inverse* of the admittance, H_{adm}^{-1}.

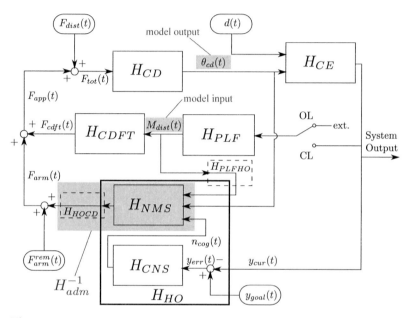

Figure 4.1: The biodynamic feedthrough system model. A human operator (HO) controls a controlled element (CE) using a control device (CD). Motion disturbances $M_{dist}(t)$ are coming from the platform (PLF). The feedthrough of $M_{dist}(t)$ to involuntary applied forces $F_{arm}(t)$ and involuntary control device deflections $\theta_{cd}(t)$ is called biodynamic feedthrough (BDFT). The feedthrough of $M_{dist}(t)$ to inertia forces $F_{cdft}(t)$ is called control device feedthrough (CDFT). $F_{app}(t)$ is the sum of the forces applied to the control device by the HO. The HO consists of a central nervous system (CNS) and a neuromuscular system (NMS). The connection between the HO and the environment is governed by two 'interfaces', H_{PLFHO} and H_{HOCD}. The CE and PLF can form an open-loop (OL) or closed-loop (CL) system. The focus of the current chapter: a physical BDFT model is proposed that models the dynamics between $M_{dist}(t)$ and $\theta_{cd}(t)$, i.e., the B2P dynamics. The model is constructed by extending an admittance model.

- The force disturbance feedthrough (FDFT) is the combination of the neuromuscular admittance and the control device dynamics, describing the transfer dynamics between force disturbance $F_{dist}(t)$ and control device deflections $\theta_{cd}(t)$.

- Several types of BDFT dynamics can be defined. The model proposed in this chapter has the $M_{dist}(t)$ signal as input and the $\theta_{cd}(t)$ signal as output. Hence, the model proposed here describes BDFT to positions (B2P) dynamics. However, as the model could be used to describe B2FCL dynamics as well, the general term biodynamic feedthrough (BDFT) model will be used throughout the chapter.

4.2.1   Force disturbance feedthrough model

Earlier studies showed that variations in the BDFT dynamics strongly correlate with variations in neuromuscular admittance (see Chapter 2). Admittance describes limb dynamics through a relationship between the forces applied to the limb and the resulting change in limb position. To describe admittance, detailed admittance models have been developed at the Delft University of Technology [Abbink, 2006; Mugge et al., 2009; Schouten et al., 2008b]. Such a model, describing the admittance of the human arm, forms the basis of the BDFT model proposed in the current chapter. The model will be introduced here, more details on the model's transfer functions will follow in Section 4.3.1.

 The elements of the admittance model are shown schematically in Fig. 4.2. The intrinsic neuromuscular dynamics of the human arm are represented by a mass-spring-damper (MSD) system, which is extended using a reflexive and muscle-activation model (not shown in full detail in the figure). Visco-elastic grip dynamics, i.e., the dynamics of the connection of the hand with the CD, are described by spring-damper dynamics. The admittance model is combined with a model for the control device, modeled through a second MSD-system, a combination referred to as the force disturbance feedthrough (FDFT) model. This model describes how the combined dynamics of the NMS and CD respond to force disturbances (see

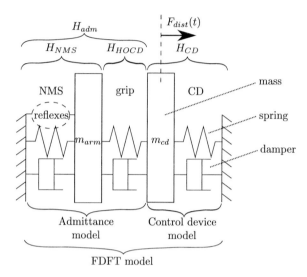

Figure 4.2: Schematic of the force disturbance feedthrough model, a combination of the admittance model (left) and control device model (right).

Chapter 3 for details on how these dynamics interact). The FDFT model describes the transfer dynamics of three of the model blocks shown in Fig. 4.1, i.e., the control device dynamics H_{CD}, the neuromuscular dynamics H_{NMS} and the interface dynamics between HO and CD, H_{HOCD}, consisting of the visco-elastic grip dynamics. An analogous FDFT model was introduced in a block representation in Fig. 3.2.

The approach taken in this study is to develop a physical BDFT model by expanding the FDFT model. The basic motivation for this is that the elements included in the FDFT model are also involved in the BDFT process. Moreover, these elements can be identified *independently* from other elements in the BDFT model, using $F_{dist}(t)$, as will be shown later.

As four different but related models will be used in the remainder of the chapter, it is helpful to summarize their relationships here:

- The *admittance model* describes neuromuscular admittance (i.e., limb dynamics) and contains parameters describing neuro-muscular, reflexive, activation and grip dynamics. As limb dynamics vary between and within subjects, also the admittance model parameters vary between and within subjects. It will be shown how the admittance model parameters can be estimated.

- The *control device model* describes the dynamics of the control device and typically consists of an MSD-system with fixed and known parameters.

- The *force disturbance feedthrough (FDFT) model* is the combination of the admittance and control device model (see Chapter 3).

- In the following, the FDFT model will be used to develop the fourth and final model: the *biodynamic feedthrough (BDFT) model*, used here to describe B2P dynamics.

4.2.2 Developing the biodynamic feedthrough model

A first step in extending the FDFT model is to allow the combination of control device and human arm to move with respect to the environment under influence of the motion disturbance $M_{dist}(t)$. This is done by placing the FDFT on a movable platform, i.e., the PLF, see Fig. 4.3. Next, a model for the interface dynamics between the platform and the human operator, H_{PLFHO}, needs to be added. The necessary elements and required complexity of these interface dynamics are subject to debate. Here only lateral motion disturbances are considered, which limits the model to describe lateral motion only. For the feedthrough of disturbances in this direction, the sideways motion of the torso (with respect to the seat) and sideways motion of the upper arm (rotation around the shoulder joint) are assumed to be the most relevant. For this BDFT model, these contributions were lumped together into one upper body model, describing the H_{PLFHO} dynamics. The structure selected for this model is an MSD system. Note that a more accurate model could

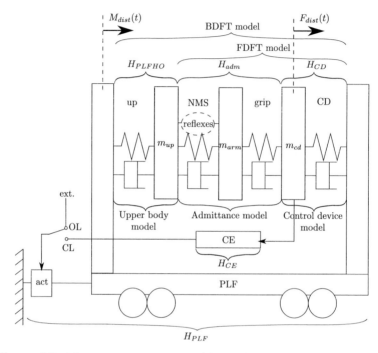

Figure 4.3: Schematic representation of the biodynamic feedthrough model, with from left to right the upper body, admittance and control device models.

be obtained by constructing an upper body model using multiple masses, representing upper-arm and torso, connected through multiple MSD systems, each extended with reflexive dynamics. One may also consider adding seat dynamics, as these dynamics play a role in how accelerations enter the human body. These additions, however, will inevitably increase the risk of over-fitting and complicate the process of estimating the model parameters correctly and reliably. An MSD system is assumed here to be the simplest dynamical structure capable of describing the expected physical process of torso and upper-arm motion. Extensions (or reductions) are possible at a later stage, when more is known about the dynamics that need to be accounted for in the H_{PLFHO} block.

Fig. 4.3 shows a schematic representation of the BDFT model; note that the original admittance and FDFT models are still recognizable on the right-hand side of the model. The upper body model, H_{PLFHO}, is shown on the left-hand side. Note that the operator's body dynamics, i.e., all MSD elements to the left of m_{cd}, vary between and (some) within subjects. The cart-like platform (PLF) can be moved by means of an actuator (act). Motion of the PLF inserts an acceleration disturbance $M_{dist}(t)$ into the system. The controlled element, CE, is also indicated and uses the CD position as input. In closed-loop BDFT systems, the CE state affects the motion of the PLF, in open-loop systems this influence is absent. This is indicated with a switch.

One dynamical element remains invisible in the current BDFT model: the operator's central nervous system H_{CNS}. The CNS is responsible for cognitive control actions, i.e., the voluntary part of manual control. In terms of the BDFT model in Fig. 4.3, the role of the CNS can be represented by varying the operator's spring, damper and reflexive dynamics, in order to simulate changes in grip, muscle co-contraction, limb movement or other voluntary effects. This was not visually represented in Fig. 4.3 to keep the figure clear, but also because this study into BDFT primarily focuses on the *involuntary* part of manual control.

4.3 Model transfer functions

4.3.1 The FDFT model

Fig. 4.4 shows the same FDFT model as presented in Fig. 4.2, but now as a block diagram. The input of the model is the force disturbance signal $F_{dist}(t)$, the output is the control device deflection $\theta_{cd}(t)$. In the following, the contents of each block will be addressed. For more information regarding the parameters and model structure see [Abbink, 2006; Mugge et al., 2009; Schouten et al., 2008b]. The block H_{cd}^I and H_{cd}^{bk} contain the inertia and spring-damper dynamics of the control device respectively:

$$H_{cd}^I(s) = 1/(I_{cd}s^2)$$

4.1

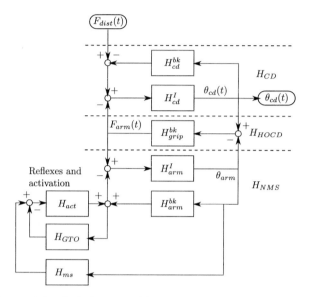

Figure 4.4: Block diagram of the FDFT model, including reflexes.

and

$$H_{cd}^{bk}(s) = k_{cd} + b_{cd}s$$

4.2

where:

- s is the Laplace variable,

- I_{cd} is the control device inertia [N m s^2 / rad]

- b_{cd} is the control device damping [N m s / rad]

- k_{cd} is the control device stiffness [N m / rad]

Note that the transfer dynamics between $F_{arm}(t)$ and $\theta_{cd}(t)$ are:

$$H_{cd}(s) = -\frac{H_{cd}^I}{1 + H_{cd}^I H_{cd}^{bk}} = -\frac{1}{I_{cd}s^2 + b_{cd}s + k_{cd}}$$

4.3

The position of the hand will slightly differ from the position of the control device. This is due to the visco-elastic properties of the skin of the hand and the grip of the hand on the control device. This effect is modeled as grip dynamics, using spring-damper dynamics, represented by H_{grip}^{bk}:

$$H_{grip}^{bk}(s) = k_{grip} + b_{grip}s \qquad \text{4.4}$$

with b_{grip} and k_{grip} the grip damping (in [N m s / rad]) and the grip stiffness (in [N m / rad]) respectively.

The arm inertia block H_{arm}^{I} contains the arm inertia, converting forces to accelerations and integrating to arm positions, $\theta_{arm}(t)$. The intrinsic stiffness and damping of the arm are captured in the block H_{arm}^{bk}. The arm dynamics are described by:

$$H_{arm}^{I}(s) = 1/(I_{arm}s^2) \qquad \text{4.5}$$

$$H_{arm}^{bk}(s) = k_{arm} + b_{arm}s \qquad \text{4.6}$$

where:

- I_{arm} is the endpoint inertia of the arm [N m s^2 / rad]

- b_{arm} is the intrinsic arm damping [N m s / rad]

- k_{arm} is the intrinsic arm stiffness [N m / rad]

The muscles are mainly controlled through commands from the CNS, traveling from the brain to the muscles. However, faster feedback loops also exist in the form of spinal reflexes. The two most important mechanisms of reflexive feedback are incorporated in the admittance model: feedback from the Golgi Tendon Organs (GTO) and feedback from the muscle spindles (ms). For an in-depth discussion of the identification and modeling of spinal reflexes, the reader is referred to [de Vlugt, 2004]. The GTOs are located where the tendon is attached to the muscle and provide force feedback. Generally, GTOs reduce overall stiffness and damping of the muscles. The GTO dynamics are incorporated in the H_{GTO} block:

$$H_{GTO}(s) = k_f e^{-s\tau_{ref}} \qquad \text{4.7}$$

with k_f the GTO feedback gain ([-]) and τ_{ref} the reflexive time delay (in [s]).

Muscle spindles are located within the muscles and provide information on muscle stretch and stretch velocity. Generally, the muscle spindles increase overall stiffness and damping of the muscles. Their dynamics are included in the H_{ms} block:

$$H_{ms}(s) = (k_p + k_v s)e^{-s\tau_{ref}}$$ 4.8

where:

- k_p is the muscle spindle stretch feedback gain [N m /rad],

- k_v is the muscle spindle stretching rate feedback gain [N m s / rad] and

- τ_{ref} is the reflexive time delay [s].

The muscle activation block H_{act} describes the process of active muscle force build-up following a neural activation signal. It is approximated by second order dynamics:

$$H_{act}(s) = 1/(\frac{1}{4\pi^2 f_{act}^2}s^2 + \frac{\beta}{\pi f_{act}}s + 1)$$ 4.9

with β ($= \frac{1}{2}\sqrt{2}$) the relative damping (see [de Vlugt, 2004]) and f_{act} the cut-off frequency (in [Hz]).

4.3.2   The BDFT model

Fig. 4.5 shows the same BDFT model as presented in Fig. 4.3, but now as a block diagram. Compare this diagram with Fig. 4.4. The blocks that were added to the FDFT model of Fig. 4.4 are indicated in gray. In the following the contents of these additional blocks will be addressed.

The upper body dynamics are modeled using lumped MSD dynamics. H_{up}^I describes the effective inertia of the upper arm and torso and H_{up}^{bk} describes the lumped stiffness and damping:

$$H_{up}^I(s) = 1/(I_{up}s^2)$$ 4.10

$$H_{up}^{bk}(s) = k_{up} + b_{up}s$$ 4.11

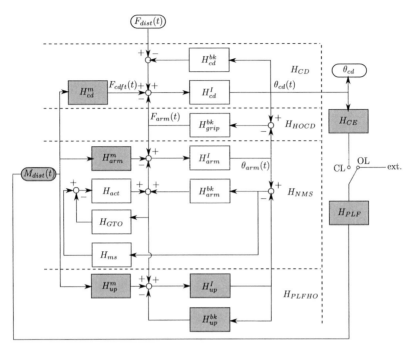

Figure 4.5: Block diagram of the BDFT model. The white model blocks are identical to those in the FDFT model in Fig. 4.4.

with I_{up}, b_{up} and k_{up} the upper body's effective inertia (in [N m s^2 / rad]), damping (in [N m s / rad]) and stiffness (in [N m / rad]) respectively.

The blocks H_{CE} and H_{PLF} represent the controlled element and platform. In case of an open-loop BDFT system, the platform acceleration signal $M_{dist}(t)$ can be regarded as an input to the model [Venrooij et al., 2013a], in effect removing the PLF dynamics from the BDFT model.

The way the acceleration signal $M_{dist}(t)$ enters the system is modeled through fictitious forces, working on each of the masses present in the system. Through elementary Newtonian dynamics ($F = ma$), these mass blocks basically convert the PLF accelerations to inertia forces that work on the center of gravity of the respective elements:

$$H_{cd}^m(s) = m_{cd}, \; H_{arm}^m(s) = m_{arm}, \; H_{up}^m(s) = m_{up} \qquad \boxed{4.12}$$

with m_{cd}, m_{arm} and m_{up} (all in [kg]) the mass of the control device, the mass of the forearm and the lumped mass of the upper body respectively. Note that the force $F_{cdft}(t)$ is the result of control device feedthrough (CDFT).

4.4   Modeling considerations

4.4.1   Open- or closed-loop

When considering the dynamics included between $M_{dist}(t)$ and $\theta_{cd}(t)$ in Fig. 4.5, one notes that the dynamics of interest do *not* include H_{CE} or H_{PLF}. Hence, a change in either or both of these systems does not affect the B2P dynamics. It is therefore reasonable to assume that the processes that underlie the B2P dynamics are the same in open-loop and closed-loop systems. Studying open-loop systems has several important experimental benefits [Venrooij et al., 2013a], such as full and exclusive control over the acceleration disturbance $M_{dist}(t)$ and the fact that the results obtained in an open-loop setting are independent from vehicle dynamics. In this study, an open-loop approach was adopted to identify and validate the proposed model, which can be applied in future closed-loop studies.

4.4.2   Two-stage parameter estimation

The main advantage of constructing a BDFT model by extending an FDFT model, as is proposed in this chapter, becomes apparent when estimating the parameter values of the model. First of all, the control device dynamics are assumed to be known: all parameters of the H_{CD} blocks in Fig. 4.5 are defined a-priori. All further parameter estimation is then performed in two stages. First, the admittance is estimated using force disturbance $F_{dist}(t)$; the estimate is used to obtain values for the admittance model parameters. After this parameter estimation step all parameters contained in the white blocks in Fig. 4.5, i.e., the FDFT model, are either known or estimated. In the second stage the B2P dynamics can be estimated with motion disturbance $M_{dist}(t)$, which allows obtaining values for the remaining parameters.

The claim that a two-stage parameter estimation is more reliable that a one-stage approach is supported with two main arguments: (i) estimating all parameters in a single step using *only* the B2P dynamics would result in large parameter value uncertainty. Especially because the model contains some repetitive elements (i.e., several masses, springs and dampers) it is difficult to obtain reliable parameter values for all parameters in a single step without putting stringent constraints on each parameter. By using the admittance in addition to the B2P dynamics, an additional sources or information becomes available for the parameter estimation. (ii) The admittance model, which formed the basis of the BDFT model proposed here, has been thoroughly studied. As a result the parameter estimation process for this model is well documented [Abbink, 2006; Mugge et al., 2009]. This is not the case for many BDFT models available in the literature, where the parameter estimation is often poorly documented or performed for only a limited number of subjects (e.g., [Jex and Magdaleno, 1978; Mayo, 1989]). Hence, the reliability of the parameter estimation is increased by incorporating knowledge on admittance modeling into the BDFT model.

It is important to note that the proposed two-stage parameter estimation method hinges on the assumption that the parameters estimated for the admittance model can be used directly in the BDFT model. As a strong interaction was found between admittance and B2P, an interaction between the respective models is also to be expected. More research is required to fully understand the limitations of this approach. For example, research has suggested that the function of the muscle spindles is relatively easily disturbed by vibrations while the GTOs are less sensitive to vibrations [Brown et al., 1967; Lewis and Griffin, 1976]. Whether and how these effects can be incorporated in the admittance model should be investigated.

4.4.3   Describing the output

There are two disturbances which can be considered as inputs: $F_{dist}(t)$ and $M_{dist}(t)$, each contributing to the output, i.e., the control device deflection signal:

$$\theta_{cd}(t) = \theta_{cd}^{Fdist}(t) + \theta_{cd}^{Mdist}(t) + \theta_{cd}^{res}(t) \qquad \boxed{4.13}$$

where superscripts ^{Fdist} and ^{Mdist} indicate the contributions of the force disturbance and of the motion disturbance to the control device deflection respectively. The *involuntary* part of the output caused by the acceleration disturbance, $\theta_{cd}^{Mdist}(t)$, is a signal of particular interest here. The third part of the output, the residual, is denoted with the superscript ^{res}, and is defined as any part of $\theta_{cd}(t)$ that is not related to one of the two disturbance signals. Note that this includes any *voluntary* cognitive contribution from the CNS and the remnant. A previous study showed that under certain circumstances this contribution remains small in comparison with the other two contributions (see Chapter 2). The FDFT model describes $\theta_{cd}^{Fdist}(t)$, the BDFT model describes B2P dynamics, i.e., the $\theta_{cd}^{Mdist}(t)$ signal. If these models provide an accurate description of the actual dynamics and if $\theta_{cd}^{res}(t)$ is indeed small, the sum of the output of the two models describes $\theta_{cd}(t)$ closely:

$$\theta_{cd}(t) \approx \bar{\theta}_{cd}(t) = \bar{\theta}_{cd}^{Fdist}(t) + \bar{\theta}_{cd}^{Mdist}(t) \qquad \boxed{4.14}$$

where $\bar{\theta}_{cd}^{Fdist}(t)$ is the output of the FDFT model and $\bar{\theta}_{cd}^{Mdist}(t)$ is the output of the BDFT model. The sum of both model outputs is the combined model output, $\bar{\theta}_{cd}(t)$, which is an approximation of the measured output $\theta_{cd}(t)$.

A straightforward approach for evaluating the combination of the FDFT and BDFT models is to calculate the difference between the measured $\theta_{cd}(t)$ and the modeled $\bar{\theta}_{cd}(t)$. To evaluate the performance of each model individually, the contributions $\theta_{cd}^{Fdist}(t)$ and $\theta_{cd}^{Mdist}(t)$ need to be obtained and compared to their modeled counterparts. The method used here to extract these signals from the measured signals is a frequency decomposition technique, as proposed in Chapter 2.

The frequency decomposition technique separates the measured signal into its contributions by splitting the total power spectral density function $PSD^{tot}(j\omega)$ (with j the imaginary unit) in the frequency domain on particular frequencies of the frequency vector ω (for details, see Section 2.4.7). The frequency decomposition results in three components of $PSD^{tot}(j\omega)$:

$$PSD^{tot}(j\omega) = PSD^{Fdist}(j\omega) + PSD^{Mdist}(j\omega) + PSD^{res}(j\omega) \quad \boxed{4.15}$$

Compare the above with Eq. 4.13, which has a similar structure, but then in the time domain. Inverse Fast Fourier transforming $PSD^{Fdist}(j\omega)$ and $PSD^{Mdist}(j\omega)$ yields time-domain signals $\theta_{cd}^{Fdist}(t)$ and $\theta_{cd}^{Mdist}(t)$. By comparing these to their modeled counterparts the quality of the FDFT and BDFT models can be evaluated individually.

4.5 Measuring neuromuscular admittance and biodynamic feedthrough

The data used to estimate the parameters of the models were obtained through an open-loop experiment in which both admittance and B2P were measured. Chapter 2 provides a detailed description and validation of the measurement approach. The experiment performed for this study was different from the one described in Chapter 2, hence the experiment description is provided here in detail. The data obtained from this experiment were also used in Chapters 3, 6 and 7.

4.5.1 Apparatus

The experiment was performed in the SIMONA Research Simulator (SRS) of TU Delft, a 6 degree-of-freedom research flight simulator [Stroosma et al., 2003]. The control device was an electrically actuated side-stick, positioned to the subject's right. Participants were sitting in an upright posture, right arm adducted and parallel to the torso. Right forearm in 90 degrees flexion, 90 degrees pronation. Digits 2-5 of the right hand were flexed anti-clockwise, digit 1 was

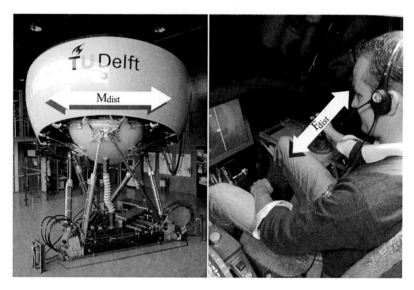

Figure 4.6: The experimental setup used in this study: a subject (right), seated inside the SIMONA Research Simulator (left) uses a side-stick to perform a control task. A force disturbance signal, $F_{dist}(t)$, perturbs the side-stick in lateral direction. A motion disturbance signal, $M_{dist}(t)$, perturbs the motion-base of the simulator in lateral direction.

flexed clockwise around the stick shaft. No armrest was present. During the experiment, a motion disturbance $M_{dist}(t)$ was applied to the simulator motion base – to measure the B2P dynamics – and a force disturbance $F_{dist}(t)$ was applied to the stick – to measure the admittance (see Fig. 4.6). The applied forces $F_{app}(t)$ and control device deflections $\theta_{cd}(t)$ were measured. In this study, only lateral (left-right) acceleration disturbances were studied. Also the neuromuscular admittance was only measured in lateral direction by using a lateral force disturbance on the control device. The control device was fixed in the longitudinal direction. A head-down display (15-in LCD, 1024×768 pixels, 60 Hz refresh rate) was located in front of the subject. The seat had a five-point safety belt that was adjusted tightly, to reduce torso motions.

Table 4.1: Data of subjects (N = 12, all male and right-handed).

	Age [years]	Weight [kg]	Height [cm]	Fore arm [cm]	Upper arm [cm]	BMI [kg/m²]
mean μ	24.3	73.4	183.6	28.3	34.3	21.9
st. dev. σ	1.9	4.2	7.2	1.9	3.5	2.1
range	21-27	68-82	174-194	25-31	29-40	18.1-24.5

4.5.2 Subjects

Fourteen subjects participated in the experiment, recruited from the TU Delft student population. The data of two subjects had to be removed due to large outliers which could only be explained by insufficient task compliance. Their data was not used for further analysis. The remaining subjects formed a homogeneous group, see Table 4.1: all subjects were male, right-handed, and had only a small variation in age (21 – 27 years) and body mass index (BMI) (18.1 – 24.5 kg/m²). The body mass index (BMI) is calculated by dividing a person's weight (in kg) by height squared (in m²), and is a measure of the total amount of body fat in adults [Maddan et al., 2008].

4.5.3 Task instruction

Subjects performed three disturbance rejection tasks (TSK), the classical tasks [Abbink et al., 2011]:

- position task (PT) (or stiff task), in which the instruction is to keep the position of the side-stick in the centered position, that is, to "resist the force perturbations as much as possible",

- force task (FT) (or compliant task), in which the instruction is to minimize the force applied to the side-stick, that is, to "yield to the force perturbations as much as possible", and

- relax task (RT), in which the instruction is to relax the arm while holding the stick, that is, to "passively yield to all side-stick perturbations".

The human operator needed to set his/her neuromuscular properties differently for optimal control of each of the three control tasks. The PT is a task for which the best performance is achieved by being very stiff (i.e., a small admittance), the FT requires the operator to be very compliant (i.e., a large admittance). The RT yields an admittance reflecting the passive dynamics of the neuromuscular system. The classical tasks have been used in numerous studies to investigate the variability of the neuromuscular system, e.g., [Abbink, 2006; Lasschuit et al., 2008; Mugge et al., 2009]. Earlier studies (see Chapter 2) indicated that the admittance, and with that the B2P dynamics, strongly depend on these control tasks.

4.5.4   Procedure

Subjects were instructed on the experiment goal and the control tasks they were to perform. Several training runs were conducted to allow subjects to get used to the disturbances and the control tasks. During training, visual performance feedback was provided on the display. A laterally moving red block displayed the parameter to be controlled and the reference was shown by a white, vertical line running down the center of the display. During the PT, the controlled parameter was the lateral side-stick deflection angle and the reference was 0 degrees; during the FT the controlled parameter was the applied force to the side-stick and the reference was 0 N. For both tasks the display also showed a time-history of position or force respectively such that subjects could monitor their performance. During the RT the display presented no information. After the execution of the task a score was provided (calculated using the standard deviation of control force and control position signal). When a consistent performance (i.e., score) was reached, the visual performance feedback was removed from the screen for the remainder of the experiment – to minimize cognitive control actions based on visual feedback – and the actual measurement started.

The measurements were performed under two conditions (COND): in the 'motion condition' (MC) both the lateral motion disturbance signal $M_{dist}(t)$ and the force disturbance $F_{dist}(t)$ were applied simultaneously. In the 'static condition' (SC) only the force disturbance $F_{dist}(t)$ was applied. Here, no biodynamic feedthrough was present and only the force disturbance response, i.e., admittance, could be measured. Together with the three control tasks, this resulted in a 3×2 repeated-measures design with two independent variables: TSK and COND. First, six repetitions of each control task in the static condition were performed, yielding admittance estimates without the effect of motion (results not shown). The tasks were performed in groups of two of the same tasks, e.g., two PT's followed by two RT's, etc. The order of the task groups was random. Second, in the motion condition, six repetitions of each control task were performed, in random order, yielding estimates of admittance and B2P dynamics. During the measurements the subject's score was provided only at the end of each run.

4.5.5 Perturbation signal design

The force disturbance $F_{dist}(t)$ was a multi-sine signal with frequency set ω_f, containing $N = 31$ frequencies ω_{f_n} ($n = 1...N$). The motion disturbance $M_{dist}(t)$ was a multi-sine signal with frequency set ω_m, containing $K = 31$ frequencies ω_{m_k} ($k = 1...K$). The sets ω_f and ω_m were chosen such that they did not overlap, see Fig. 4.7. This allows the response due to each disturbance to be identified in the measured signals [Abbink, 2006; Venrooij et al., 2011a]. To obtain a full bandwidth estimate of the admittance, a range of frequencies between 0.05 Hz and 21.5 Hz were selected for ω_f. For the BDFT estimate, a range of frequencies between 0.1 and 21.5 Hz was selected for ω_m.

4.5.6 Perturbation signal scaling

To minimize the effect of non-linearities and to be able to compare between different control tasks, the stick deflection in each task should be small and similar in size. In an effort to achieve this, the disturbance signals were scaled in a tuning procedure [Venrooij et al., 2011a]. The goal of this procedure was to find the gains

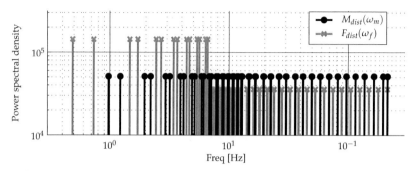

Figure 4.7: Power spectral density plot of disturbance signals $F_{dist}(t)$ and $M_{dist}(t)$.

needed to result in similar standard deviation (3 degrees) of the control device deflection $\theta_{cd}(t)$ for each control task. The scaling factors for $F_{dist}(t)$ found for the PT and FT were 20 and 1 respectively, resulting in signal root-mean-square (RMS) values of the $F_{dist}(t)$ signal of 10.17 N and 0.67 N. During the tuning process it was observed that the admittance estimate of the RT seemed to depend on the gain of the force disturbance signal. In order to minimize the gain effect, the requirement of 3 degrees standard deviation was ignored for the RT and instead, the minimal disturbance gain was determined at which a reliable estimate was possible at an acceptable coherence level. This gain was found to be 0.38 (with an $F_{dist}(t)$ RMS of 0.39 N). For the $M_{dist}(t)$ signal, the scaling factors (RMS of $M_{dist}(t)$ in parentheses) for PT, FT and RT were 0.9 (0.79 m/s^2), 0.7 (0.62 m/s^2) and 0.7 (0.62 m/s^2) respectively.

4.5.7   Non-parametric identification

The admittance was estimated in the frequency domain, using the estimated cross-spectral densities between $F_{dist}(t)$ and $\theta_{cd}(t)$ ($\hat{S}_{fdist,\theta}(j\omega_f)$) and between $F_{dist}(t)$ and $F_{arm}(t)$ ($\hat{S}_{fdist,f}(j\omega_f)$) [van der Helm et al., 2002] (for more details on this procedure see Chapter 3):

$$\hat{H}_{adm}(j\omega_f) = \frac{\hat{S}_{fdist,\theta}(j\omega_f)}{\hat{S}_{fdist,f}(j\omega_f)}$$

4.16

with j the imaginary unit. This procedure assumes linearity, and to check the reliability of this assumption, the squared coherence – a measure for the signal-to-noise ratio (SNR) with a value between 0 and 1 – was calculated [van der Helm et al., 2002]:

$$\hat{\Gamma}^2_{adm}(j\omega_f) = \frac{\left|\hat{S}_{fdist,\theta}(j\omega_f)\right|^2}{\hat{S}_{fdist,fdist}(j\omega_f)\hat{S}_{\theta,\theta}(j\omega_f)} \qquad \text{4.17}$$

The frequency response describing the B2P dynamics can be estimated using the estimated cross-spectral density between $M_{dist}(t)$ and $\theta_{cd}(t)$ ($\hat{S}_{mdist,\theta}(j\omega_m)$)) and the estimated auto-spectral density of $M_{dist}(t)$ ($\hat{S}_{mdist,mdist}(j\omega_m)$)):

$$\hat{H}_{B2P}(j\omega_m) = \frac{\hat{S}_{mdist,\theta}(j\omega_m)}{\hat{S}_{mdist,mdist}(j\omega_m)} \qquad \text{4.18}$$

The corresponding squared coherence function equals:

$$\hat{\Gamma}^2_{B2P}(j\omega_m) = \frac{\left|\hat{S}_{mdist,\theta}(j\omega_m)\right|^2}{\hat{S}_{mdist,mdist}(j\omega_m)\hat{S}_{\theta,\theta}(j\omega_m)} \qquad \text{4.19}$$

4.5.8 Experimental data

Estimates of neuromuscular admittance (measured in MC) and B2P dynamics for a typical subject are shown in Fig. 4.8. The admittance clearly depends on the control task; as was reported earlier, it is the lowest in the position task and the highest in the force task [Abbink, 2006; Mugge et al., 2009; Venrooij et al., 2011a]. Furthermore, the B2P dynamics clearly differ for each control task as well [Venrooij et al., 2011a]. Fig. 4.9 shows the squared coherences of the admittance and B2P estimates. The high squared coherences obtained for the B2P estimate indicate a reliable estimate of the dynamics. The squared coherences for the admittance estimate are lower, e.g., around 1 Hz and for the RT, but still acceptable for the overall frequency range.

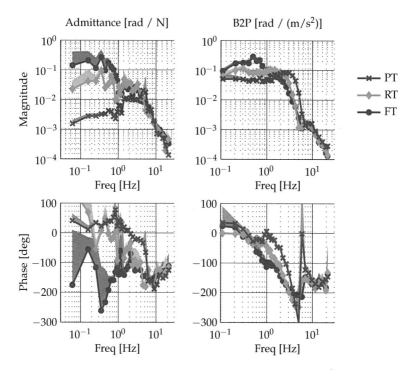

Figure 4.8: Estimates of the neuromuscular admittance ($\hat{H}_{adm}$) and B2P ($\hat{H}_{B2P}$) for a typical subject. Means over repetitions (lines) and 1 standard deviation (colored bands) are shown.

4.6 Parameter estimation procedure

4.6.1 The admittance model

The parameters of the admittance model were estimated per subject, by fitting the admittance model on the estimate of the admittance $\hat{H}_{adm}(j\omega_f)$ for each control task. Recall that the admittance model is the part of the FDFT model that describes the admittance, see Fig. 4.2. After the parameter estimation of the admittance model, the FDFT model follows directly by combining it with the control device dynamics [Venrooij et al., 2013a].

The reliability of the parameter estimation was evaluated using the

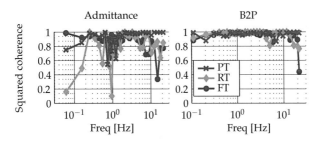

Figure 4.9: Squared coherence of the admittance and B2P estimate of a typical subject.

standard error of the mean (SEM). The SEM is a measure of the variance of the parameter distribution. Generally, parameters that have little contribution to the prediction error, show large variances and a large standard error of the mean [Abbink, 2006; de Vlugt et al., 2006; Ljung, 1999].

The admittance model describes the admittance by ten parameters, see Table 4.2. Most of these parameters were expected to vary within subjects (across control tasks) due to adaptations of the neuromuscular system. Three parameters, I_{arm}, τ_{ref} and f_{act}, were assumed constant or not to vary significantly across control tasks, therefore these parameters were kept constant within a given subject (they were allowed to vary between subjects, though). The muscle-stretch feedback gain k_p was assumed to be zero in all cases, as this parameter showed a large SEM, indicating it could not be estimated reliably. This was also observed and dealt with in this way in other studies, where it was found that the parameter had a negligible contribution to the error criterion and the value of other parameters did not change much with the absence or presence of k_p [Abbink, 2006; Mugge et al., 2009].

The above procedure leads to 6 parameters that vary per subject and task and 3 parameters that vary per subject, resulting in 21 parameters ($6 \times 3 + 3$) to be estimated per subject. To guide the fit and to prevent unrealistic parameter values and bad convergence, boundaries were set on each parameter, see Table 4.2. For each subject the admittance parameters were fit for the three control tasks simultaneously, by minimizing the sum of the squared logarithmic difference in admittance (using MATLAB's `lsqnonlin` function), with the following error criterion:

$$E_{adm} = \sum_{TSK} \sum_{\omega_f} \left| \log \left[\frac{\hat{H}_{adm}(TSK, \omega_f)}{\bar{H}_{adm}(TSK, \omega_f)} \right] \right| \qquad \boxed{4.20}$$

with TSK denoting the control tasks (PT, FT, RT) and ω_f the frequencies of the force disturbance signal. $\hat{H}_{adm}$ is the frequency response function (FRF) of the admittance estimate and $\bar{H}_{adm}$ is the FRF of the admittance model. Note that the value of E_{adm} reflects the difference between the model and the obtained data over a range of frequencies and is a metric for the quality of the frequency domain fit.

4.6.2   The BDFT model

The BDFT model requires 5 parameters in addition to the known CD parameters and the parameters that were estimated for the admittance model in the previous step, see Table 4.2. Relying on well-documented procedures for the parameter estimation of the admittance model to obtain 13 of the 18 parameters of the BDFT model increases the reliability of the parameter estimation procedure as a whole. This two-stage parameter estimation is a major advantage of the modeling approach proposed here.

Three constraints were imposed on the parameter values to guide the parameter estimation procedure to reasonable values. First, it was assumed that the parameters I_{up}, m_{up} and m_{arm}, i.e., the masses and inertias, do not to vary within a subject. They were kept constant across tasks for a given subject, i.e. they were allowed to vary only between subjects. Second, physical principles dictate that a relationship exists between the masses (m_{up} and m_{arm}) and their respective inertias (I_{up} and I_{arm}). Here, it was assumed that:

$$m_{arm} = p_{arm} I_{arm} \text{ and } m_{up} = p_{up} I_{up} \qquad \boxed{4.21}$$

with p a scalar (unit [rad/m^2]) that relates inertia to mass. Note that I_{arm} already followed from the parameter estimation of the

Table 4.2: Model parameters and parameter bounds.

Control device model			
Parameter	Unit	Lower bound	Upper bound
Control device dynamics			
1.1[b] I_{cd}	[N m s^2 / rad]	set to 0.1770	
1.2[b] b_{cd}	[N m s / rad]	set to 1.3587	
1.3[b] k_{cd}	[N m / rad]	set to 11.6926	
Admittance model			
Grip dynamics			
2.1 b_{grip}	[N m s / rad]	0.0	100
2.2 k_{grip}	[N m / rad]	0.0	5000
Intrinsic arm dynamics			
2.3[a] I_{arm}	[N m s^2 / rad]	0.05	0.3
2.4 b_{arm}	[N m s / rad]	0.1	7
2.5 k_{arm}	[N m / rad]	0.1	500
Golgi tendon organ feedback			
2.6 k_f	[-]	-20	20
Muscle spindle feedback			
2.7 k_v	[-]	0	50
2.8[b] k_p	[-]	set to 0.0 [Abbink, 2006]	
Time delay			
2.9[a] τ_{ref}	[s]	0.005	0.05
Activation dynamics			
2.10[a] f_{act}	[Hz]	0.5	3.0
(Additional parameters of the) BDFT model			
Upper-body dynamics			
3.1[b] I_{up}	[N m s^2 / rad]	set to 2.089	
3.2 b_{up}	[N m s / rad]	0	60
3.3 k_{up}	[N m / rad]	0	1000
Effective masses			
3.4[a] p_{arm} (Eq. 4.21)	[rad / m^2]	0	20
3.5[a] p_{up} (Eq. 4.21)	[rad / m^2]	0	20

[a] Parameter fixed within subject
[b] Parameter fixed for all subjects and conditions

admittance model. As it can be expected that the mass/inertia relationship is similar across subjects and independent of the inertia or mass, estimating p instead of m effectively removes one subject-dependent variable.

One final parameter constraint was included, which followed from observations after the first BDFT model parameter estimation attempts. Parameters I_{up}, b_{up} and k_{up} showed large SEMs, indicating that the model was not sensitive to changes in these parameters, possibly because of over-parameterization. It was observed that the value I_{up} did not vary widely across subjects (μ=2.089, σ=0.404 Nms^2/rad), possibly due to the subject group homogeneity (see Table 4.1). Fixing the value I_{up} for all subjects also greatly reduced SEMs (improving the reliability of the parameter estimation), without compromising the model quality in the frequency or time domain (this was checked). Hence, I_{up} was set to the mean value obtained across subjects, being 2.089 Nms^2/rad.

The observation that the MSD dynamics used as the upper body model showed large SEMs, which could be reduced by fixing the value of I_{up}, may raise some doubts on how adequate this model structure is. As was already indicated in Section 4.2.2, an MSD structure was chosen here to be the simplest dynamical structure capable of describing the expected physical process of torso and upper-arm motion. More complex models may be more adequate and closer to reality. Extension of the complexity of the upper body model was *not* pursued in the current study, however, for two main reasons: firstly, extending the upper body model without proper knowledge on which dynamics are relevant to include severely increases the risk of over-parameterization, a problem plaguing many physical models. Secondly, the current BDFT model serves primarily the purpose of increasing the understanding of the relationship between neuromuscular admittance and BDFT. An extensive upper body model would inevitably degrade the role of the admittance model in the overall dynamics. In fact, a rich upper body model, with many parameters, could 'cover up' inaccuracies in the admittance model and obscure the true relationship between admittance and BDFT. In other words, a detailed upper body model would be capable of describing BDFT dynamics, regardless of whether the admittance model is correct or even relevant in the occurrence of BDFT. Conversely, *if* an admittance model can be converted to a BDFT model using such an elementary upper body model as used here, it can be interpreted as evidence that a strong relationship between admittance and BDFT exists.

The above-mentioned constraints on the BDFT model parameters resulted in two parameters to be estimated per task and subject (b_{up} and k_{up}), and two only per subject (p_{up} and p_{arm}), which resulted in 8 parameters ($2 \times 3 + 2$) to be estimated for every subject. The error criterion to be minimized in the BDFT model parameter estimation was:

$$E_{BDFT} = \sum_{TSK} \sum_{\omega_m} \left| \log \left[\frac{\hat{H}_{B2P}(TSK, \omega_m)}{\bar{H}_{BDFT}(TSK, \omega_m)} \right] \right| \qquad \text{4.22}$$

with ω_m the frequencies of the motion disturbance signal. $\hat{H}_{B2P}$ is the FRF of the B2P estimate and $\bar{H}_{BDFT}$ is the FRF of the BDFT model.

4.7 Analysis in the time domain

Minimizing error criterion Eq. 4.22 yields a fit of the BDFT model on the measured data. The minimum value of E_{BDFT} reflects the quality of the fit in the frequency domain. Another, equally important and common evaluation of the model quality can be obtained in the time domain: the variance-accounted-for (VAF) [Abbink, 2006]. It reflects how well the variance of a measured signal is approximated by its simulated counterpart. For the FDFT model the VAF is calculated using the difference between the measured contribution of $F_{dist}(t)$ to the CD deflection, $\theta_{cd}^{F_{dist}}(t)$, and the modeled signal, $\bar{\theta}_{cd}^{F_{dist}}(t)$:

$$VAF_{FDFT} = \left[1 - \frac{\sum_{k=1}^{N} \left[\theta_{cd}^{F_{dist}}(t_k) - \bar{\theta}_{cd}^{F_{dist}}(t_k) \right]^2}{\sum_{k=1}^{N} \left[\theta_{cd}^{F_{dist}}(t_k) \right]^2} \right] \qquad \text{4.23}$$

with N denoting the number of data points in the time domain. Note that when $\bar{\theta}_{cd}^{F_{dist}}(t)$ is equal to $\theta_{cd}^{F_{dist}}(t)$, the VAF value is 100%. The time signal $\bar{\theta}_{cd}^{F_{dist}}(t)$ can be obtained from simulation by providing the force disturbance as input to the FDFT model. Time-signal $\theta_{cd}^{F_{dist}}(t)$ can be obtained by using the frequency decomposition technique described in Chapter 2.

The VAF for the BDFT model fit was calculated from the difference between the measured contribution of $M_{dist}(t)$ to the CD deflection, $\theta_{cd}^{M_{dist}}(t)$, and the modeled signal, $\bar{\theta}_{cd}^{M_{dist}}(t)$:

$$VAF_{BDFT} = \left[1 - \frac{\sum_{k=1}^{N} \left[\theta_{cd}^{M_{dist}}(t_k) - \bar{\theta}_{cd}^{M_{dist}}(t_k) \right]^2}{\sum_{k=1}^{M} \left[\theta_{cd}^{M_{dist}}(t_k) \right]^2} \right] \qquad 4.24$$

Finally, also the combination of the FDFT and BDFT models can be evaluated by calculating the VAF of the combined model output $\bar{\theta}_{cd}$, as defined in Eq. 4.14:

$$VAF_{comb} = \left[1 - \frac{\sum_{k=1}^{N} \left[\theta_{cd}(t_k) - \bar{\theta}_{cd}(t_k) \right]^2}{\sum_{k=1}^{N} \left[\theta_{cd}(t_k) \right]^2} \right] \qquad 4.25$$

4.8   Results

4.8.1   The admittance model

Fig. 4.10 shows a typical frequency-domain fit of the admittance model. The modeled dynamics approximate the non-parametric admittance estimates well, in magnitude and phase. Results of the time-domain model validation, using the VAF (Eq. 4.23), are shown in Table 4.3. The values represent the averaged VAF values obtained across all subjects for each task. The table indicates mean and standard deviation (σ). The VAF values obtained in the PT and FT are above 80%, evidence for good model quality. For the relax task the VAF is considerably lower, around 42%. This is largely due to the small force disturbance gain used in this task (Sec. 4.5.6), to mitigate a 'gain effect' observed for the RT in earlier work [Venrooij

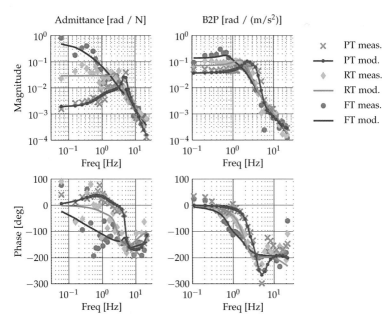

Figure 4.10: Fit of the admittance model on the admittance data (left) and the BDFT model on the B2P data (right) for a typical subject.

et al., 2013a]. Lowering the disturbance gain, however, inevitably results in a reduction in the signal-to-noise ratio (SNR), leading, amongst other effects, to a lower squared coherence, see Fig. 4.9. Lower SNRs also lead to reduced VAF values, as a larger portion of the measured signals is noise, not accounted for in the model. Comparing a noisy measurement signal with its modeled counterpart – which should be a noise-free approximation – will show a low VAF, even when a perfect model is used. For lower SNRs the VAF becomes a less reliable metric for model quality. In the frequency domain it can be observed that the model is the best possible smooth approximation of the (noisy) measured data. It should be acknowledged that the SNR of the data obtained for this task is low, reducing the certainty with which the admittance parameters for the RT model can be estimated. However, it should also be

Table 4.3: Average Variance Accounted For (VAF).

	VAF_{FDFT} (FDFT model)	VAF_{BDFT} (BDFT model)	VAF_{comb} combined
Task	VAF (σ) [%]	VAF (σ) [%]	VAF (σ) [%]
PT	83.92 (5.8)	82.75 (6.3)	79.63 (6.1)
RT	42.49 (10.7)	88.48 (3.8)	83.78 (3.8)
FT	82.20 (8.0)	87.29 (3.7)	85.39 (4.1)

noted that in this case, the VAF is a rather unreliable metric and its low value does not necessarily imply an inadequate model (see also [Venrooij et al., 2014b]).

The parameter values for a typical subject are shown in Fig. 4.11 (excluding k_p, as it was fixed to 0.0 for all conditions). Note that the parameters I_{arm}, f_{act} and τ_{ref} (indicated with white bars) were kept fixed across tasks, but were still allowed to vary across subjects. Furthermore, note that the delay time τ_{ref} shows a large SEM, indicating that it could not be estimated reliably. The highest intrinsic stiffness k_{arm} and grip stiffness k_{grip} were obtained during the PT, the 'stiff' task where a high stiffness and strong grip is useful, confirming earlier work [Abbink, 2006; Mugge et al., 2009]. The lowest stiffness values are found for the FT, which agrees with the 'compliant' task instruction. For this task a relatively high amount of reflexive behavior was found, a finding that also agrees with previous research. The reflex activity in the relax task is found to be very small. A small negative value for the GTO feedback k_f is found in the position task, as reported as well for the ankle joint in [Abbink, 2006; Mugge et al., 2009]. It has been suggested that a high stiffness setting is reflexively supported by negative GTO activity, instead of high muscle spindle activity [Abbink, 2006; Mugge et al., 2009]. Hence, the results obtained here are in full agreement with what has been reported in the literature. More can be said about the admittance parameters, but the focus currently lies with the BDFT model.

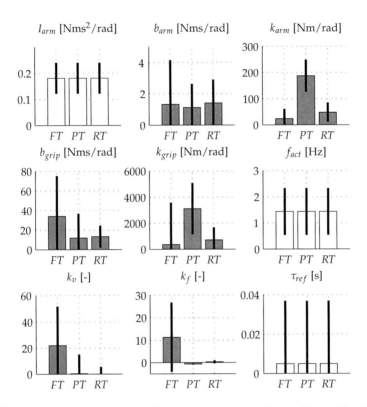

Figure 4.11: Admittance model parameters for a typical subject. The bars indicate the parameter values, the lines indicate the estimated standard errors of the mean (SEM). The parameters that were fixed across control tasks are indicated in white.

4.8.2 The BDFT model

Fig. 4.10 shows a BDFT model fit on the measurement data, indicating a good approximation in both magnitude and phase, for each task. Table 4.3 lists the VAF values obtained for the BDFT model under VAF_{BDFT}: high VAF values were obtained, across tasks, and the small standard deviations indicate that similar results were obtained across subjects. Together with the adequate fit in the frequency domain, this provides strong evidence that the BDFT

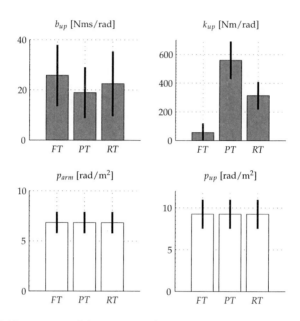

Figure 4.12: BDFT model parameters for a typical subject. The bars indicate the parameter values, the lines indicate the estimated standard errors of the mean (SEM). The parameters that were fixed across control tasks are indicated in white.

model as proposed in this chapter, including the proposed method of parameter estimation, allows for accurate BDFT modeling across different subjects (accounting for between-subject variability) and across control tasks (accounting for within-subject variability), in frequency and time domain.

The parameter values of the BDFT model (excluding I_{up} as it was fixed for all conditions), obtained for a typical subject, in addition to the ones in Fig. 4.11, are shown in Fig. 4.12. It can be seen that p_{arm} and p_{up} were held constant for the three control tasks and the parameters b_{up} and k_{up} vary per control task. The low SEMs indicate the reliability of the parameter estimation. The PT shows a high stiffness, the FT a low stiffness, the RT stiffness lies in-between, all in full agreement with expectations regarding the body dynamics for each control task.

The results also provide insight in the physical background of the BDFT phenomena, one of the main reasons to develop a physical model. For example, the PT shows high stiffness (k_{arm}, k_{grip} and k_{up}) combined with low damping (b_{arm}, b_{grip} and b_{up}). Typically, a high stiffness and low damping reduce the stability of a mass-spring-damper system; this might explain the BDFT 'peak' occurring around 2-3 Hz for the PT. The FT shows a low stiffness and high damping, removing the instability but also making the system more vulnerable to low-frequency disturbances; and indeed, at low frequencies a high B2P magnitude can be clearly observed in the B2P dynamics measured for this task. Also note that k_{up} of the BDFT model and k_{arm} of the admittance model seem to share features across the different control tasks. The same can be said for b_{up} and b_{arm}. This is important, as it might suggest that these parameters are related. Regarding the physical system that is modeled, i.e., a set of connected limbs, such a relationship would agree with expectations, as changing admittance is known to involve many muscles across the forearm, upper arm and torso [Damveld et al., 2010].

The VAF obtained for the combined model output, VAF_{comb}, is shown in Table 4.3 under VAF_{comb}. The combination of the two models provides an accurate description of the total control device deflection. This implies that the residual component $\theta_{cd}^{res}(t)$ of Eq. 4.13, which is not accounted for in the model and includes cognitive contributions of the CNS, is small compared to the contributions of the two disturbances.

4.8.3 Results across subjects

Fig. 4.13 shows the *measured* B2P dynamics (magnitude only) for the 12 individual subjects used in this study, separated for each control task, as thin gray lines. The average B2P dynamics, obtained by averaging the measured data of the individual subjects, is also shown (thick black line). These results show that the B2P dynamics vary across the conditions, i.e., *within* subjects, and also differ *between*

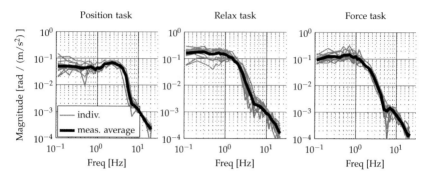

Figure 4.13: Magnitude of *measured* B2P dynamics for each individual subject and the average of the *measured* B2P dynamics over all subjects.

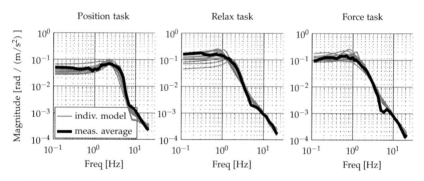

Figure 4.14: Magnitude of the *modeled* B2P dynamics for each individual subject and the average of the *measured* B2P dynamics over all subjects.

subjects (e.g., the RT data show a considerable spread). The BDFT model, proposed in the current chapter should be able to account for both between- and within-subject variability.

In Fig. 4.14 the measured individual data has been replaced by the *modeled* B2P dynamics for each subject. For reference purposes, the same average measured dynamics (thick black line) is repeated in this figure. The comparison of Fig. 4.13 and Fig. 4.14 provides insight in the quality of the BDFT model across different subjects in general terms. It can be observed that the models reproduce many features of the measured dynamics and approximate the average in

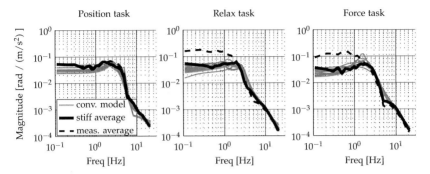

Figure 4.15: Magnitude of *modeled* B2P dynamics for each individual subject, converted to stiff CD dynamics and the average of the *measured* B2P dynamics for the stiff CD dynamics. The measured average of the original data (with normal CD dynamics) is also shown (dashed line).

a similar way, e.g., the spread in the measured dynamics for the RT is reproduced by the associated models. Some model weaknesses can be observed as well. It can be observed that the model yields slightly lower B2P dynamics than was actually measured for the low frequencies of the PT. Furthermore, the 'dent' that is typically present in the measured B2P dynamics between 5-7 Hz is not reproduced by the model (particularly clear for the FT). This may be due to the very simple upper body model used in the current model. More complete upper body models, e.g., including seat dynamics, may improve this. Despite these potential weaknesses, it can be concluded from the comparison that the BDFT model proposed here is generally capable of providing good approximation of the measured dynamics and incorporates both between- and within-subject variability.

4.8.4   A sanity check: adapting control device dynamics

Chapter 3 provides a detailed discussion on how a change in control device dynamics influences the B2P dynamics. It was shown that by making the control device dynamics stiffer (by changing the CD stiffness from 0.2 N/deg to 1.2 N/deg), the B2P magnitude for the FT and RT decreases, but remains unchanged for the PT condition. These changes were explained using the interactions between FDFT, B2FOL and B2P dynamics (see page 107).

In the following, this adaptation of CD dynamics is reproduced in the BDFT model developed in the current study. A successful 'conversion' of the modeled dynamics would indicate that the influence of the control device dynamics is correctly represented in the model. Fig. 4.15 shows the results of this conversion operation. The figure indicates the converted model dynamics for each subject (i.e., only k_{cd} was changed from 0.2 N/deg to 1.2 N/deg, no further changes were made) as thin gray lines. The average of the *measured* data obtained for the *stiff* CD dynamics is also shown (solid black line). For reference purposes, the original measured average (the same as in Figs. 4.13 and 4.14) is also shown (dashed black line).

The results show that the modeled dynamics change in agreement with expectations. The converted models approximate the measured stiff average closely. As expected, a large reduction can be observed for the RT and FT magnitude, while the PT magnitude remains virtually unchanged. It should also be noted that for the PT the modeled dynamics again show an underestimation of the actual dynamics, just as was observed in Fig. 4.14. That is, the magnitudes of the converted models are lower than the measured stiff average for low frequencies. The exact cause for this underestimation, which is evidence of an inaccuracy in the model, is yet unknown and requires further investigation. In general, however, it can be concluded that the conversion was successful, indicating that the influence of control device dynamics is correctly represented in the model. This provides additional evidence that the model structure indeed describes the physical phenomena it aims to reproduce.

4.9 Discussion

Just as any other model, the physical BDFT model proposed here is useful in some cases and less useful in others. The current model was developed as a physical BDFT model, based on neuromuscular principles. The model provides novel insights in several of the influencing factors that play a role in BDFT. Factors like the control device dynamics, seat dynamics or upper body dynamics can be scrutinized in unprecedented detail. Furthermore, the fact that the model is so strongly dependent on admittance sheds light on the important relationship between admittance and BDFT [Venrooij et al., 2010a, 2011b]. However, there are also some disadvantages related to the model: estimation of the parameters of the model requires measuring the admittance of the human operator, which may be hard to do outside the laboratory. Measuring the admittance as was done in this study, using a force disturbance, is most likely not practical in an actual vehicle. Furthermore, the parameter estimation is elaborate, requires several assumptions and does not estimate all parameters accurately.

Hence, if one requires a model providing physical insight in the BDFT phenomenon, allowing for a detailed investigation of its influencing factors, this model may be very valuable. If one, on the other hand, requires an efficient and tangible model, describing BDFT at an end-point level, without requiring a physical interpretation of each parameter, this model is probably not a good choice. Regarding future work with the current model: it is worthwhile to investigate whether it is possible to *predict* the BDFT model parameter values from the admittance model parameters. This would be an important step forward in BDFT research. The challenge that would then remain is obtaining an (online) estimate of the admittance. Another aspect that deserves attention is the upper body model, as more complete models may improve model quality and provide additional insights.

While constructing the model and estimating its parameters, several assumptions were made and constraints were imposed because of the limited knowledge available on the neuromuscular principles that govern BDFT. After all, this model is the first of its kind and prior to constructing the model much was still unknown. This holds in particular for the upper-body dynamics, which were modeled here as a mass-spring-damper system of which the inertia I_{up} was fixed to one value across all conditions. It is highly likely that different choices will lead to better results than those obtained and presented here. This was not pursued in this thesis, as research efforts were devoted to an alternative mathematical model instead, which will be addressed in the next chapter. The model presented in the current chapter should therefore be interpreted as a promising first iteration which requires further research efforts to come to full fruition.

4.10 Conclusions

A biodynamic feedthrough model was proposed and developed by extending an admittance model to incorporate control device dynamics and to account for the effect of accelerations. In this way, a physical BDFT model is obtained, which serves primarily the purpose of increasing the understanding of the relationship between admittance and biodynamic feedthrough. One of the major contributions of this model is its capability to describe both between-subject and within-subject BDFT variability from physical principles, something that is not included in existing BDFT models. The model parameters can be used to gain insight in the physical background of the BDFT phenomena, e.g., explaining how muscle stiffness and damping influence the measured behavior. An added advantage of the proposed modeling approach is that the model parameters can be estimated using a two-stage approach, which makes the parameter estimation more robust, as it is largely based on the well-documented procedure used for the admittance model. The data used to estimate the model parameters were obtained through an open-loop experiment in which both admittance and B2P were measured. The parameters obtained for the admittance model agreed with the results obtained in earlier studies. The quality of the models was evaluated in frequency and time domain and the results provide strong evidence that the BDFT model proposed in this chapter, including the proposed method of parameterizing it, allows for accurate BDFT modeling across different subjects and across control tasks. The time domain analysis showed average values of Variance Accounted For (VAF) around 80% across different conditions, signifying the model provides an accurate description for different conditions tested in this study. By changing the control device stiffness and comparing the model results with measurement data, it was confirmed that the influence of the control device dynamics on B2P are correctly represented by the model.

CHAPTER 5

A mathematical biodynamic feedthrough model for rotorcraft

This chapter presents a mathematical biodynamic feed-through model for rotorcraft. This model aims to fill the gap between currently existing simple but therefore often limited black box models and the versatile but more complex physical models. The model structure was obtained through asymptote modeling, which offers a structural method to design a model's transfer function. The resulting model was thoroughly evaluated in both the frequency and the time domain. Furthermore, the model's performance was compared to two black box models and a physical model. The results of the validation show that the mathematical BDFT model provides a highly accurate description of BDFT dynamics.

The contents of this chapter are based on:

Paper title   A Mathematical Biodynamic Feedthrough Applied to Rotorcraft

Authors   Joost Venrooij, Mark Mulder, David A. Abbink, Marinus M. van Paassen, Max Mulder, Frans C. T. van der Helm, and Heinrich H. Bülthoff

Published in   IEEE Transactions on Cybernetics, 2013, Vol. 44, No. 7, July 2014

5.1 Introduction

Biodynamic feedthrough (BDFT) is recognized as a problem for rotorcraft for a variety of operations [Gabel and Wilson, 1968; Mayo, 1989; Walden, 2007] and handling qualities are known to degrade due to BDFT effects [Rodchenko et al., 1993]. The occurrence of BDFT in rotorcraft has been under investigation for several decades. Recent studies have been conducted in the context of the GARTEUR HC-AG16 project (e.g., [Dieterich et al., 2008]) and the ARISTO-TEL project (e.g., [Masarati et al., 2013; Pavel et al., 2012; Quaranta et al., 2013; Venrooij et al., 2011c]). These projects mainly investigate Rotorcraft-Pilot Couplings (RPCs), i.e., oscillations or divergent vehicle responses, originating from adverse pilot-vehicle couplings. Biodynamic feedthrough can both cause and sustain such events.

The recent interest in RPCs is driven by the fact that the rapid implementation of more advanced flight control systems appears to have caused more RPC events than before [Pavel et al., 2011, 2012]. This stresses the need to obtain a more fundamental understanding of how BDFT may interfere with control performance in order to predict, evaluate and alleviate its effects on RPCs.

However, the complexities of BDFT, and especially the present limited understanding of BDFT in general, have kept researchers and manufacturers in the rotorcraft domain from developing robust ways of dealing with BDFT (and RPCs). An important and necessary step forward is accurate *modeling* of the BDFT phenomenon. A BDFT model can not only be used to gain understanding of the phenomenon, it can also provide ways towards an effective and robust solution of the problem. Such a model is the topic of the current chapter.

It should be noted that, in addition to rotorcraft, many other vehicles suffer from BDFT problems as well and that also for these cases the accurate modeling of BDFT can be equally valuable. In comparison, however, the interest in and efforts directed into BDFT research seems larger in the rotorcraft community than in many other areas. This indicates, first of all, that BDFT is of particular importance for this vehicle type and, secondly, that there is a demand for BDFT models devoted to rotorcraft, such as the one that is presented in this chapter.

In general, the *methods and analysis* proposed in this chapter are directly applicable to other vehicle types. Due to differences in vehicle dynamics, experimental design details (e.g., disturbance signal design) may differ between studies, but the approach to model BDFT is generally transferable. It is not possible, however, to transfer BDFT *models* directly, as BDFT is strongly dependent on vehicle and control device dynamics. In this study a model was developed using data obtained in a helicopter setup, confining the scope of application of the model to this vehicle type. If one is interested in developing a similar BDFT model for, e.g., fixed-wing aircraft, the experiment described in this chapter could be repeated, using a setup resembling an aircraft cockpit with appropriate control devices.

Several BDFT models have been developed in the past decades, but only a few of them account for variability between or within subjects. For rotorcraft, a simple BDFT model was proposed in [Mayo, 1989], incorporating between-subject variability by providing different parameter sets for two different body types. In [Venrooij et al., 2013b] this model was extended by providing parameter sets for three different body types and for different settings of the neuromuscular system, including both between-subject and within-subject variability. It was noted that in some conditions the model was unable to describe BDFT effects at higher frequencies, which highlights one of the possible weaknesses of a simple model structure: it may not be suitable to describe all relevant dynamics.

More versatile – and thus more complex – BDFT models are also available, e.g., [Jex and Magdaleno, 1978; Venrooij et al., 2014a] (see also Chapter 4). These physical models may include muscle dynamics, limb masses, reflex activity etc. These models typically provide and accurate description of the BDFT dynamics and allow for a detailed physical understanding of the phenomenon [Venrooij et al., 2014a]. A drawback of these models is that they are usually complex and their parameters are often difficult to estimate, rendering them unwieldy and hard to use in practical situations [Venrooij et al., 2013b].

Simply put, two types of BDFT models exist in current literature:

easy-to-use limited ones and hard-to-use extended ones. The model proposed in the current chapter aims to fill this gap. The BDFT model that is proposed here is capable of describing important BDFT features, measured across subjects and in different conditions, using a simple transfer function structure which can be implemented easily. The model is applicable to rotorcraft and offers an accurate description for BDFT effects for three different motion directions. The method used to obtain the model structure, asymptote modeling, is described in detail and can be applied to other modeling problems as well. In the current chapter, suitable parameter sets for this model structure will be proposed and validated. Using these parameters, the model can be directly implemented in many typical rotorcraft BDFT studies.

The chapter is structured as follows: Section 5.2 discusses BDFT modeling and different model types. Section 5.3 describes how the experimental data was obtained on which the model is based. Section 5.4 elaborates on how a model structure was chosen and how its parameters were estimated. Extensive model validation results are discussed in Section 5.5. The chapter ends with conclusions and recommendations in Section 5.6.

5.2 Modeling biodynamic feedthrough

5.2.1 The biodynamic feedthrough system model

The BDFT system model, shown in Fig. 5.1, helps to gain an understanding of the elements that play a role in biodynamic feedthrough. It contains all the high-level elements of a general BDFT system. Chapter 3 provides an elaborate discussion of the BDFT system model, which will not be repeated here. The BDFT system model is shown here for convenience and to indicate the input and output of the mathematical BDFT model developed in the current chapter.

Not all the elements of the BDFT system model will be included in the mathematical BDFT model. The mathematical model aims to describe only the transfer dynamics between the motion disturbances that enter the human body (input) and the involuntary

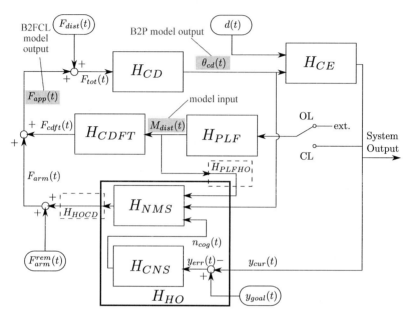

Figure 5.1: The biodynamic feedthrough system model. A human operator (HO) controls a controlled element (CE) using a control device (CD). Motion disturbances $M_{dist}(t)$ are coming from the platform (PLF). The feedthrough of $M_{dist}(t)$ to involuntary applied forces $F_{arm}(t)$ and involuntary control device deflections $\theta_{cd}(t)$ is called biodynamic feedthrough (BDFT). The feedthrough of $M_{dist}(t)$ to inertia forces $F_{cdft}(t)$ is called control device feedthrough (CDFT). $F_{app}(t)$ is the sum of the forces applied to the control device by the HO. The HO consists of a central nervous system (CNS) and a neuromuscular system (NMS). The connection between the HO and the environment is governed by two 'interfaces', H_{PLFHO} and H_{HOCD}. The CE and PLF can form an open-loop (OL) or closed-loop (CL) system. The focus of the current chapter: a mathematical BDFT model is developed, which describes B2FCL$^+$ dynamics, i.e., the dynamics between $M_{dist}(t)$ and $F_{app}(t)$. By multiplying the model with the CD dynamics, a B2P model is obtained.

forces applied to the control device (output), see Fig. 5.1. This implies the model developed in this chapter describes biodynamic feedthrough to forces in closed-loop (B2FCL) dynamics.

5.2.2 Relevant dynamics from the BDFT system model

Biodynamic feedthrough models aim to describe how vehicle accelerations $M_{dist}(t)$ cause involuntary forces $F_{arm}(t)$ and involuntary control device deflections $\theta_{cd}(t)$. By following the path from $M_{dist}(t)$ to $F_{arm}(t)$ and $\theta_{cd}(t)$ in Fig. 5.1, it can be observed that the following elements play a role in modeling BDFT dynamics: the neuromuscular system H_{NMS}, the control device H_{CD} and the two interfaces H_{PLFHO} and H_{HOCD}. Note that only the control device has invariant (and usually known) dynamics, all other elements can vary between and within subjects.

Furthermore, note that the above list of influencing dynamics does not include the H_{CE}, H_{PLF} and H_{CNS} blocks. The first two blocks describe vehicle dynamics, which are relevant to describe the acceleration signals the HO is exposed to, and – in closed-loop systems – the vehicle motion in response to BDFT induced inputs. These blocks are therefore of interest when studying, e.g., the stability of the human-vehicle interaction [Quaranta et al., 2013], however, they do not influence the BDFT dynamics themselves. It is reasonable to assume that the processes that underly BDFT are the same in open-loop and closed-loop systems [Venrooij et al., 2013a]. This allows for studying BDFT in an open-loop fashion, i.e., exposing a subject to a predetermined acceleration signal $M_{dist}(t)$ and measuring the forces and control device deflections, without having to 'close the loop' by feeding these inputs into a vehicle model.

Although many (but certainly not all) practical BDFT problems occur in closed-loop cases, studying the open-loop system has several important experimental benefits [Venrooij et al., 2013a]. A prime advantage is the experimenter's full and exclusive control over the acceleration disturbance $M_{dist}(t)$. Secondly, the results obtained in an open-loop setting are independent from vehicle dynamics, and hence the resulting model is more generic (in our case that means the model is generally applicable to rotorcraft, not one rotorcraft type in particular). In this study, an open-loop approach was adopted to construct and validate a BDFT model, which can be applied in future closed-loop studies.

The CNS is responsible for cognitive control inputs, which are of great importance when modeling voluntary control behavior. As biodynamic feedthrough is strictly involuntary in nature, the contents of this block are not influencing the occurrence of BDFT directly. Indirectly, there are two main ways the CNS influences BDFT dynamics.

First of all, the pilot can cognitively correct for the involuntary BDFT induced control inputs. As it is known that the cognitive bandwidth is usually limited to frequencies below 1 Hz [Allen et al., 1973], it can be assumed that the CNS influences the BDFT dynamics primarily below this frequency [Quaranta et al., 2013]. In Chapter 2 it was shown through frequency decomposition that for experimental conditions similar to the ones used in this chapter, the influence of cognitive corrective control actions is relatively small.

The second way the CNS can influence the BDFT dynamics is by commanding adaptations of the neuromuscular system, such as modulation of the reflexive behavior and muscle co-contraction. These modulations influence the BDFT dynamics for a larger frequency range, including higher frequencies. The experimental design used in this study controls for this influence by asking the subjects to perform classical tasks that require a particular setting of the neuromuscular system. In doing so, the effect of cognitive neuromuscular adaptation on BDFT dynamics can be systematically studied. Note that understanding these effects does not require to model the CNS itself, as the neuromuscular adaptation is expressed through changes in the NMS.

5.2.3 Biodynamic feedthrough to forces and positions

The effects of BDFT manifest themselves as involuntary *forces*, i.e., an involuntary component in $F_{arm}(t)$, and involuntary control device *positions*, i.e., an involuntary component in $\theta_{cd}(t)$. To distinguish between these two simultaneously occurring effects, it was proposed in Chapter 3 to label them 'biodynamic feedthrough to forces' (abbreviated as B2F) and 'biodynamic feedthrough to positions' (abbreviated as B2P) respectively. B2P refers to the transfer dynamics between the measurable signals $M_{dist}(t)$ and $\theta_{cd}(t)$. BDFT to forces in closed-loop (B2FCL) refers to the transfer dynamics between the measurable signals $M_{dist}(t)$ and $F_{app}(t)$. From Fig. 5.1 follows a relationship between B2P and B2FCL[a]:

$$H_{B2P}(s) = H^+_{B2FCL}(s)H_{CD}(s)$$

5.1

where $H_{CD}(s)$ are the control device dynamics (see also Eq. 3.49 on page 94).

Recall that the B2FCL$^+$ dynamics are the force equivalent of the B2P dynamics. In the following, a B2FCL$^+$ model will be developed that is converted to a B2P model using Eq. 5.1. A major advantage of using B2FCL$^+$ for constructing models, instead of using B2P directly, is that it is only weakly depending on the control device dynamics, as opposed to B2P. By multiplying the B2FCL$^+$ model with the control device dynamics of choice, one obtains an approximation of the desired B2P dynamics. However, it should be noted that B2FCL$^+$ is not fully independent from control device dynamics. Using the model and its parameters proposed here with control device settings other then those used in this study (listed in Table 5.1) may result in model inaccuracies. To construct a model that is completely independent from control device dynamics, more involved techniques are required where, e.g., the neuromuscular admittance needs to be measured (see Chapter 4).

5.2.4   Physical, black box and mathematical models

In general, two types of BDFT models are in use: physical BDFT models and black box BDFT models. Both aim to describe BDFT dynamics, but through different modeling approaches. Physical models are geared towards providing a physical representation of the

---

[a]Note that this relationship makes use of so-called 'uncorrected' B2FCL dynamics, indicated with a superscripted $^+$. The dynamics are not corrected for CDFT dynamics. As the final goal is to create a B2P model, correcting the B2FCL dynamics is not required, see Chapter 3.

BDFT phenomenon, using a-priori knowledge and physical principles (e.g., [Hess, 1998; Jex and Magdaleno, 1978; Sirouspour and Salcudean, 2003; Venrooij et al., 2014a]). This type of BDFT models typically provides additional insight in the physical processes underlying the BDFT phenomenon. In Chapter 4 a physical model was
proposed that allows investigating the influence of neuromuscular
properties, like muscle stiffness or reflexive activity, on the BDFT
dynamics. This additional insight, however, comes at a price: describing the complexities of reality calls for complex models, which
are often intrinsically over-parameterized. This can make the parameter estimation challenging or even impossible [Venrooij et al.,
2014a].

In contrast, black box models aim to provide an efficient BDFT
description at 'end-point level' (e.g., [Mayo, 1989; Sövényi, 2005]).
They do not aim to describe actual physical phenomena and are
therefore often more efficient and easier to use compared to their
physical counterparts. A comprehensive discussion on physical
and black box BDFT models is provided in Section 2.3 of [Sövényi,
2005] (p. 15).

As the current aim is to propose a practical model, to be used in
the rotorcraft domain, a black box model seems to be the appropriate choice. A black box model protects researchers and manufacturers from some of the complexities of the BDFT phenomena
(like the influence of the modulation of reflexive behavior), which
are often not their direct interest and outside their field of exper-
tise. However, black box models come with several disadvantages
as well. Due to the lack of a physical interpretation of the model
structure and its parameters, selecting and evaluating them is difficult. How does one determine the required model order? How
does one evaluate parameter values or the particular influence of a
model parameter on the dynamics? Often, black box model structures are determined using a trial-and-error approach: different
model structures, based on engineering judgment, are applied to
measured data and a combination of performance metrics determines the 'optimal model structure'.

A more systematic approach to developing the model structure
is proposed in this chapter. In Section 5.4 it will be shown that

through this approach an efficient model is obtained, in which each parameter has a distinctive influence and a bounded range of allowable values. This means the black box is no longer completely black, as the role of each parameter is clearly defined and changing a parameter value has a predictable result. However, the parameters do not have a physical interpretation either, and therefore the model is strictly speaking neither a physical nor a black box model. A suitable term might be a mathematical model, as instead of a physical interpretation the parameters have a mathematical interpretation. Models where a structure is chosen – based on insight and/or measurement data – before performing the identification are sometimes referred to as gray box models. The mathematical model proposed here would fit into this category. Here, the term mathematical model will be used to stress the important property that the structure is chosen such that the model parameters retain their mathematical interpretation, something that is not necessarily the case for all gray box models.

5.3 Obtaining experimental data

Experimental data were obtained using a method based on the method elaborately addressed in Chapter 2. For this study use was made of rotorcraft control devices (cyclic and collective). This section provides a detailed of the experiment.

5.3.1 Experimental design

Apparatus

The experiment was performed on the SIMONA Research Simulator of Delft University of Technology, a six degree-of-freedom flight simulator [Stroosma et al., 2003]. The control devices were electrically actuated cyclic and collective controls with adjustable dynamics settings. The settings used for each control axis were based on the rotorcraft handling qualities research experiments conducted by Mitchell et al. [1992] and are listed in Table 5.1. The settings of these control devices were kept constant during the experiment.

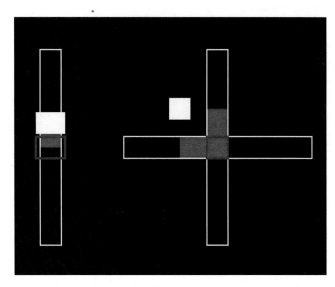

Figure 5.2: Display presented to the subject. The white outlines show the axes of the control devices. On the left the collective, on the right the cyclic (roll in horizontal and pitch in vertical direction). The white blocks indicate the position of the sticks, blue squares show the target position. Red bars along the sticks' axes help the subject to see the difference between target and actual position. The figure shows the collective slightly above target value of 50%. The cyclic is slightly deflected to the left and forward.

Friction or other non-linearities were not included in the control device dynamics. A helicopter seat was used, in which the subjects were strapped-in with a 5-point safety belt. Performance information was displayed on a head-down display (15-in LCD, 1024×768 pixels, 60 Hz refresh rate) in front of the subject. Fig. 5.2 shows the information presented to the subject.

Subjects

Fourteen right-handed subjects participated (11 males, 3 females). The subjects were volunteers from the Delft University of Technology. The data of two subjects had to be removed due to large outliers which could only be explained by insufficient task compliance.

Table 5.1: Control device dynamical settings.

Axis	Inertia [Ns^2/deg]	Damping [Ns/deg]	Stiffness [N/deg]	Length [mm]
Cyclic pitch	0.0369	0.0514	1.8340	575
Cyclic roll	0.0162	0.0516	1.8100	650
Collective	0.0152	0.0447	1.7950	600

Table 5.2 lists the subject data of the remaining twelve subjects. The body mass index (BMI) is calculated by dividing a person's weight (in kg) by height squared (in m^2), and is a measure of the total amount of body fat in adults [Maddan et al., 2008].

Experiment design

In the experiment, subjects were asked to hold the control devices and perform different disturbance rejection tasks. Two disturbance signals were used simultaneously: an acceleration disturbance $M_{dist}(t)$, applied to the simulator, and a force disturbance $F_{dist}(t)$, applied to the control devices. Using the acceleration disturbance $M_{dist}(t)$, the BDFT dynamics were determined; force disturbance $F_{dist}(t)$ permitted obtaining the neuromuscular admittance, i.e., limb dynamics [Venrooij et al., 2011a]. Motion disturbance $M_{dist}(t)$ consisted of a translational acceleration signal, applied to a single axis of the simulator. Force disturbance $F_{dist}(t)$ consisted of a force signal, applied to a single axis of the control device. The directions of $M_{dist}(t)$ and $F_{dist}(t)$ were always aligned. Measurements were performed for three motion disturbance directions (DIR): (i) lateral (LAT), (ii) longitudinal (LNG) and (iii) vertical (VRT). The force disturbance was applied to the roll axis of the cyclic in the LAT condition, to the pitch axis of the cyclic for the LNG condition and to the collective for the VRT condition.

Subjects were instructed to perform three disturbance rejection tasks (TSK) [Abbink, 2006]: (i) position task (PT), or 'stiff task', in which the instruction was to keep the position of the control devices centered, that is, to "resist the force perturbations as much as possible"; (ii) force task (FT), or 'compliant task', in which the instruction

Table 5.2: Data of subjects (N = 12).

	Age [years]	Weight [kg]	Height [cm]	BMI [kg/m^2]
mean	27.9	75.0	179.9	23.1
st. dev.	4.3	12.2	6.5	2.8
range	23-38	58-105	167-190	19.9-29.1

was to minimize the force applied to the control devices, that is, to "yield to the force perturbations as much as possible"; and (iii) relax task (RT), in which the instruction was to relax the arm while holding the control devices, that is, to "passively yield to all perturbations". For the PT the best performance is achieved by being very stiff, the FT requires the operator to be very compliant. The RT yields an admittance reflecting the passive dynamics of the neuromuscular system. Earlier studies showed that the BDFT dynamics strongly depend on these control tasks (see Chapter 2).

Each task was trained before the experiment started. The three tasks combined with the three directions results in a 3×3 repeated-measures design, each condition was repeated 6 times. During the experiment the angular deflection of the control device $\theta_{cd}(t)$ and the applied force to the control device $F_{app}(t)$ were measured.

Disturbance signal design

Both disturbance signals, $F_{dist}(t)$ and $M_{dist}(t)$, were multi-sines, defined by their frequency components. The signals were separated in frequency to allow distinguishing the response due to each disturbance in the measured signals (see Chapter 2). The frequency contents of the disturbance signals were equal in all conditions. Their magnitude was varied in such a way that the standard deviations of the control device deflections were approximately similar in each condition, to allow comparison across conditions [Venrooij et al., 2011a]. To obtain a full bandwidth estimate of the admittance, a range between 0.05 Hz and 21.5 Hz was selected for the force disturbance signal F_{dist}. This frequency range will be referred to as ω_f. For the motion disturbance signal $M_{dist}(t)$, a range between 0.1 and

21.5 Hz was selected, referred to as ω_m. For ω_f 31 logarithmically-spaced frequency points were selected in the frequency range, for ω_m 36 frequency points were selected (see [Venrooij et al., 2011c] for details). There existed no overlap between ω_f and ω_m.

5.3.2 Analysis

The B2P dynamics are estimated using the estimated cross-spectral density between $M_{dist}(t)$ and $\theta_{cd}(t)$ ($\hat{S}_{mdist,\theta}(j\omega_m)$) and the estimated auto-spectral density of $M_{dist}(t)$ ($\hat{S}_{mdist,mdist}(j\omega_m)$):

$$\hat{H}_{B2P}(j\omega_m) = \frac{\hat{S}_{mdist,\theta}(j\omega_m)}{\hat{S}_{mdist,mdist}(j\omega_m)} \qquad \text{5.2}$$

where $\hat{H}_{B2P}(j\omega_m)$ is the estimate of the actual B2P dynamics $H_{B2P}(j\omega)$ on frequencies ω_m.

The squared coherence, used to investigate the reliability of the estimate, is:

$$\hat{\Gamma}^2_{B2P}(j\omega_m) = \frac{\left|\hat{S}_{mdist,\theta}(j\omega_m)\right|^2}{\hat{S}_{mdist,mdist}(j\omega_m)\hat{S}_{\theta,\theta}(j\omega_m)} \qquad \text{5.3}$$

The B2FCL$^+$ dynamics are calculated in a very similar way, but now using the estimated cross-spectral density between $M_{dist}(t)$ and $F_{app}(t)$ ($\hat{S}_{mdist,f}(j\omega_m)$) and the estimated auto-spectral density of $M_{dist}(t)$ ($\hat{S}_{mdist,mdist}(j\omega_m)$):

$$\hat{H}^+_{B2FCL}(j\omega_m) = \frac{\hat{S}_{mdist,f}(j\omega_m)}{\hat{S}_{mdist,mdist}(j\omega_m)} \qquad \text{5.4}$$

where $\hat{H}^+_{B2FCL}(j\omega_m)$ is the estimate of the actual B2FCL$^+$ dynamics $H^+_{B2FCL}(j\omega)$ on frequencies ω_m. Note that by using the $F_{app}(t)$ signal, and not $F_{arm}(t)$, the $\hat{H}^+_{B2FCL}(j\omega_m)$ dynamics are not corrected for CDFT dynamics. As the dynamics are meant to be used to construct a B2P model (using Eq. 5.6) such a correction is not required. The squared coherence for the B2FCL$^+$ dynamics is:

$$\hat{\Gamma}^2_{B2FCL}(j\omega_m) = \frac{\left|\hat{S}_{mdist,f}(j\omega_m)\right|^2}{\hat{S}_{mdist,mdist}(j\omega_m)\hat{S}_{f,f}(j\omega_m)} \qquad \text{5.5}$$

The estimates $\hat{H}_{B2P}(j\omega_m)$ and $\hat{H}^+_{B2FCL}(j\omega_m)$ are related in the same way as in Eq. 5.1:

$$\hat{H}_{B2P}(j\omega_m) = \hat{H}^+_{B2FCL}(j\omega_m) H_{CD}(j\omega_m) \qquad \boxed{5.6}$$

From a model describing $H^+_{B2FCL}(j\omega)$, based on the estimate $\hat{H}^+_{B2FCL}(j\omega_m)$, a model describing $H_{B2P}(j\omega)$ can be easily obtained by multiplying it with the control device dynamics.

The neuromuscular admittance was also estimated using $F_{dist}(t)$, (see [Venrooij et al., 2011a,c] for details), but as the results of the admittance analysis will not be discussed further in this chapter, no further details on that analysis will be provided (see, e.g., Chapter 2 instead).

5.3.3   Results

Fig. 5.3 shows the magnitude and phase of the B2P dynamics, averaged over all subjects, for each condition, grouped per disturbance direction. The means over the subjects are indicated by the lines, the standard deviations (SD) by the colored bands (mean + 1 SD). These dynamics indicate the level of feedthrough of accelerations into involuntary control device deflections. It can be observed that the B2P dynamics depend on both disturbance direction and task. More particularly, for all three directions, for disturbances above 1-2 Hz, the PT results in the highest level of B2P. For this task, also a peak in the B2P level can be observed at approximately 2-3 Hz for each direction. This implies that 'stiff' behavior, although mostly beneficial at lower frequencies, is the worst strategy when dealing with motion disturbances above 1-2 Hz.

Fig. 5.4 shows the magnitude and phase of the B2FCL$^+$ dynamics, the level of feedthrough of accelerations into involuntary applied forces. As before, the results were obtained by averaging over all subjects, for each condition. Recall that B2FCL$^+$ data are directly related to the B2P data by multiplying them with the control device dynamics (Eq. 5.6), which are identical for all conditions in this experiment. The effect of this multiplication is that the B2P dynamics follows the second-order dynamics of the control device, for this experiment resulting in a decay in magnitude above 2-3 Hz.

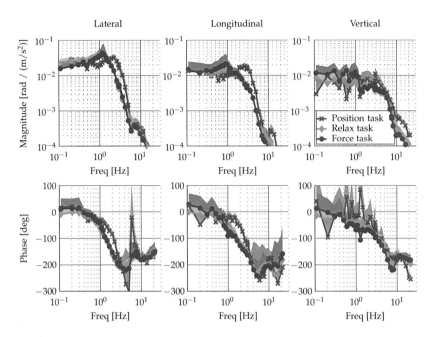

Figure 5.3: B2P dynamics – magnitude and phase – per direction and per task. Means over repetitions (lines) and 1 standard deviation (colored bands) are shown.

The absence of this decay in the B2FCL$^+$ dynamics makes it easier to observe the differences that exist between the tasks and directions (do note the difference between the y-axes between Figs. 5.3 and 5.4). For the B2FCL$^+$ dynamics it can be even more clearly observed that the dynamics depend on both disturbance direction and task. The peak occurring for the PT stands out clearly in the B2FCL$^+$ data.

Fig. 5.5 shows the squared coherence for each condition, grouped per direction. The coherence found in the lateral direction is close to 1 for each frequency, indicating a reliable estimate was obtained. The coherences for the longitudinal direction are somewhat lower, especially for the PT, but are still regarded as acceptable. However,

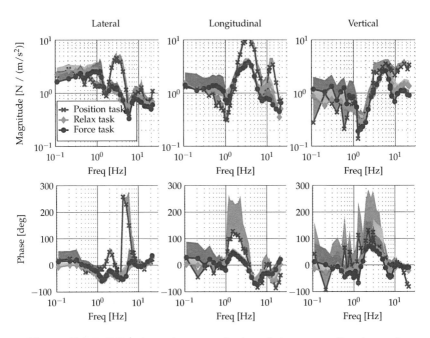

Figure 5.4: B2FCL$^+$ dynamics – magnitude and phase – per direction and per task. Means over repetitions (lines) and 1 standard deviation (colored bands) are shown.

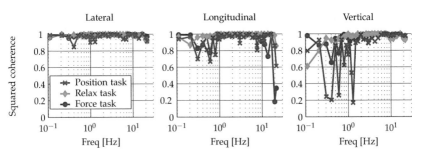

Figure 5.5: Squared coherence of the BDFT measurements in each direction.

looking at the squared coherence for the PT in the vertical direction we see very low values, especially for the lower frequencies. The cause of this is most probably the limited motion space of the SIMONA simulator in the vertical direction (1.2 m, 60% w.r.t. the lateral and longitudinal direction). Although a near maximum disturbance magnitude was used, the perturbations were apparently insufficient to obtain a high coherence between input and output. Also, the data in this direction were measured with the collective stick, and not with the cyclic, which was used for the two other directions. Therefore, both the control device dynamics and the posture of the arm for the vertical direction differ from those in the lateral and longitudinal directions. These factors may have influenced the measurement as well. The fact that the measurement results for the vertical direction are of poor quality is rather unfortunate since the BDFT effects for this direction are of particular interest [Gabel and Wilson, 1968; Venrooij et al., 2011c]. The results for all directions will be shown in the following, but it should be kept in mind that the results in the vertical direction are based on poor data and serve exploratory purposes only. Future studies should aim at obtaining data with higher coherences for the vertical direction, in order to provide more conclusive results.

5.4   Model development

In this section a method will be proposed to systematically develop a mathematical model to describe the measured B2FCL$^+$ dynamics shown in Fig. 5.4.

5.4.1   Asymptote modeling

In the approach proposed here it is assumed that *measured dynamics* can be approximated – in the frequency domain – by combining several simple functions, called *base functions*. The combination of these functions form the *modeled dynamics*, which should approximate the measured dynamics. Each base function has particular features which can be tuned to minimize the difference between measured and modeled dynamics. One feature of the base function is of particular interest: its asymptotic behavior, i.e., the behavior for lower and higher frequencies. The base functions will be tuned such that their asymptotic behavior matches a part of the measured dynamics. Hence, the approach is referred to as asymptote modeling.

To illustrate asymptote modeling it will be applied to the B2FCL$^+$ dynamics obtained for the relax task in the lateral direction. It will be shown later that it can be applied to the data obtained for the other tasks and directions as well.

5.4.2   The base functions

First the base function type is determined. Depending on the measured dynamics, one or more different types of base function might be used. For reasons that will become apparent later, here the following base function was selected:

$$H_B(s, \omega_n, \zeta, \gamma) = \left(1 + 2\zeta/\omega_n s + s^2/\omega_n{}^2\right)^\gamma \qquad \boxed{5.7}$$

where s is the Laplace variable, ω_n the natural frequency [rad], ζ the damping factor [-] and γ the order [-]. Note that in case $\gamma = -1$, the base function can be written as:

$$H_B = \frac{1}{1 + \frac{2\zeta}{\omega_n}s + \frac{1}{\omega_n{}^2}s^2} = \frac{\omega_n{}^2}{\omega_n{}^2 + 2\zeta\omega_n s + s^2} \qquad \boxed{5.8}$$

which can be recognized as a typical second-order mass-spring-damper (MSD) system. In the frequency domain, the magnitude of such a system for frequencies well below the natural frequency – its low frequency asymptote (LFA) – is flat, with a value of 1. Its high frequency asymptote (HFA), for frequencies well above the natural frequency, is a downwards slope of -40 dB/decade. For the phase, the LFA is 0° and the HFA is -180°.

It may be more convenient here to express the natural frequency in [Hz] instead of [rad], as this is also the unit used in Figs. 5.3 and 5.4. By converting [rad] to [Hz] one obtains:

$$H_B(s, f_n, \zeta, \gamma) = \left(1 + 2\zeta/(2\pi f_n s) + s^2/(2\pi f_n)^2\right)^\gamma \qquad \boxed{5.9}$$

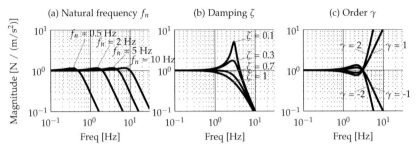

Figure 5.6: By tuning the natural frequency f_n, the damping factor ζ and the order γ, the base function changes shape.

where f_n is the natural frequency in Hz.

Fig. 5.6(a) shows the influence of the natural frequency on the magnitude of the dynamics (the phase is not discussed here). The transition from the LFA to the HFA section is determined by the damping factor ζ: reducing ζ yields a peak in the magnitude, see Fig. 5.6(b). The direction and steepness of the HFA can be determined by the order γ. For $\gamma = 1$, the HFA has a positive slope of +40 dB/decade, for $\gamma = 2$ the slope becomes +80 dB/decade, etc. Fig. 5.6(c) illustrates the influence of the order on the slope of the basis function. A final possible adaptation to the overall magnitude of the base functions (not shown in Fig. 5.6) can be made by multiplying H_B with a gain, K, with which the whole function shifts up, for $K > 1$, or down, for $K < 1$.

5.4.3 Combining base functions

The measured dynamics can be approximated by H_{mod} through multiplying a gain and several base functions:

$$H_{mod} = KH_B^1 H_B^2 H_B^3 H_B^4...$$

5.10

When multiplying base functions, the following simple rule of thumb applies: the magnitude and phase of H_{mod} is the sum of that of the base functions. For example, the combination of two base functions with a HFA of -40 dB/decade yields a HFA of -80 dB/decade; a

HFA of -40 dB/decade combined with one of +80 dB/decade results in +40 dB/decade; etc. As the base functions have a 'flat' LFA they have no influence on the combined dynamics below their natural frequency ω_n. It is the steepness of the upward or downward HFA, governed by order γ, that largely shapes dynamics of H_{mod}, together with the damping term ζ, which shapes the transition between LFA and HFA for each base function. These three properties make the second-order system an appropriate choice as base function, at least in this case. Note that if one would use a first order system, this would not allow for shaping the transition between LFA and HFA, due to the absence of a damping term. Technically, higher order systems may also be suitable, or even required, as base functions. In general, however, simpler base functions are to be preferred over complex ones, as the point of using base functions is to build up complex dynamics from simple elements. The choice of the right base function, or combination of base functions, largely depends on the dynamics that need to be modeled.

By combining base functions with different natural frequencies, damping factors and orders, the measured dynamics can be approximated. Generally, each base function accounts for a *change in slope* in the magnitude of the measured dynamics. Therefore, the measured dynamics can be split into several regions according to changes in the slope, see Fig. 5.7. There are 5 changes in slope in the measured dynamics, the model will be constructed using 5 base functions. It was aimed to develop a high-fidelity model up to 10 Hz, as it can be assumed that any (involuntary) input above this frequency will not lead to significant vehicle responses and are therefore of limited practical importance.

5.4.4 Determining the orders

The orders of the base functions can be determined by the direction and steepness of the slopes of the sections indicated in Fig. 5.7. For example, region B has a slope of approximately -40 dB/decade, region C has a slope of +40 dB/decade. Two base functions can describe this behavior when the orders of the base functions are chosen to be $\gamma_1 = -1$ and $\gamma_2 = +2$. This choice results in a HFA

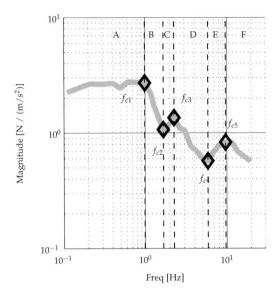

Figure 5.7: The measured dynamics can be separated in regions according to changes in slope.

of -40 dB/decade and +80 dB/decade for the two base functions respectively. Combined, this yields a -40 dB/decade slope between f_{c1} and f_{c2}, and a +40 dB/decade slope onwards. Along the same lines of reasoning the orders of the other base functions can be determined, resulting in:

$$\gamma_1 = -1, \quad \gamma_2 = +2, \quad \gamma_3 = -2, \quad \gamma_4 = +2 \text{ and } \gamma_5 = -1$$

From observation it follows that these orders apply also to the other control tasks, even for other directions. Hence, they will be kept at these values for all conditions.

5.4.5 Determining the natural frequency and damping

What remains to be determined are the natural frequencies and damping factors. Note that the desired natural frequencies are close to the frequencies where the slope changes that are shown

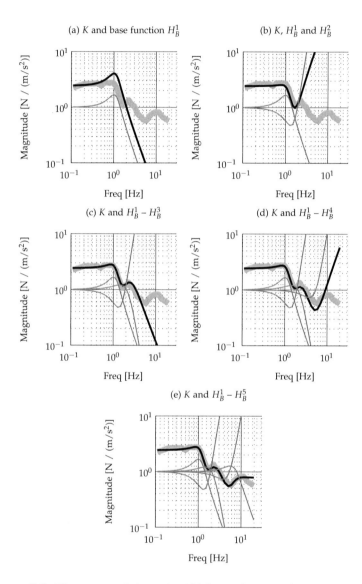

Figure 5.8: The measured dynamics (thick gray line) can be approximated by adding multiple base functions. The thin gray lines are the individual base functions, the thick black line is the modeled dynamics, i.e., the product of gain and base functions.

in Fig. 5.7 occur. The damping factors influence the 'sharpness' of a slope change. Determining the natural frequencies and damping factors could be done manually, but better results are obtained if this is done using an optimization algorithm which selects the parameter values by minimizing the difference between the measured and modeled dynamics. The following objective error criterion was used:

$$E_{B2FCL} = \sum_{TSK} \sum_{\omega_m} \left| \log \left[\frac{\hat{H}_{B2FCL}^+(TSK, \omega_m)}{\bar{H}_{mod}(TSK, \omega_m)} \right] \right| \qquad \boxed{5.11}$$

with TSK denoting the control tasks (PT, FT, RT) and ω_m the frequencies of the motion disturbance signal. The error criterion provides a measure for the difference between the measured dynamics $\hat{H}_{B2FCL}^+$ and the modeled dynamics $\bar{H}_{mod}$ for each task and frequency.

The results of this process are illustrated in Fig. 5.8, each sub-figure illustrating the effects of adding one extra base function. Fig 5.8(e) shows the final result of asymptote modeling on the B2FCL$^+$ data for the relax task in the lateral direction. Note that parameter estimation was done for all base functions simultaneously, in Fig. 5.8 the separate base functions are added in succession only to show the individual contribution of each of them to the modeled dynamics.

5.4.6 Results

The resulting models for each direction and control task are shown in Fig. 5.9. Note that the measured dynamics presented here are the same as shown in Fig. 5.4. The model approximates the measured dynamics well, for both magnitude and phase. The model parameters found for each base function are listed in Table 5.3. As the values for the orders γ_1, γ_2, etc., were fixed to the same values for all conditions, they are omitted from the table. Note that the natural frequencies are indeed close to the approximate natural frequencies identified in Fig. 5.7. Furthermore, note that the damping value for each base function is between 0 and 1. In total, the model developed in this chapter requires 16 parameters (5 base functions

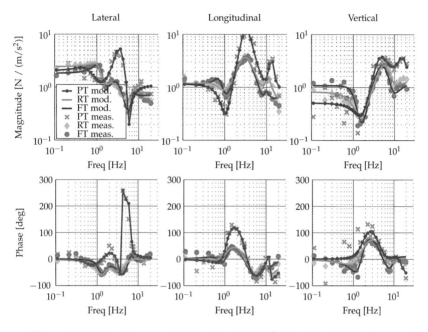

Figure 5.9: Resulting model fit of the B2FCL$^+$ data for all directions and control tasks.

× 3 function parameters + 1 gain). In many other cases it would be difficult to reliably fit 16 parameters on the measured data. However, as each parameter has a distinctive role in the model their values can be determined easily and with a relatively high degree of certainty. Just as parameters in a physical model have a physical interpretation, the parameters in the proposed mathematical model have a 'mathematical interpretation'. The orders could already be fixed before the parameter estimation by observing the slopes of the measured dynamics; the natural frequencies could be properly bounded, based on the frequencies where slope changes occur and finally, by bounding the damping ratio between 0 and 1, the parameter freedom is limited, allowing for a reliable parameter estimation.

The reliability of the parameter estimation can be evaluated using

Table 5.3: Model parameters of the global scope model.

$$H_{mod} = K H_B^1 H_B^2 H_B^3 H_B^4 H_B^5$$

with $H_B^1 = \left(1 + 2\zeta_1/(2\pi f_{n1}s) + s^2/(2\pi f_{n1})^2\right)^{\gamma_1}$, etc.

and $\gamma_1 = -1$, $\gamma_2 = +2$, $\gamma_3 = -2$, $\gamma_4 = +2$ and $\gamma_5 = -1$

	Lateral			Longitudinal			Vertical		
	FT	PT	RT	FT	PT	RT	FT	PT	RT
K [-]	1.83	2.03	2.43	1.12	1.14	1.17	1.07	0.50	0.82
f_{n1} [Hz]	1.11	0.61	1.12	0.76	0.86	0.71	1.00	1.39	0.96
f_{n2} [Hz]	1.64	1.36	1.58	1.01	1.06	1.05	1.60	1.65	1.60
f_{n3} [Hz]	2.81	3.65	2.22	3.09	3.21	3.08	5.08	4.28	5.61
f_{n4} [Hz]	6.10	6.38	5.22	9.42	10.91	9.98	8.71	10.91	8.56
f_{n5} [Hz]	7.22	6.80	7.02	9.95	12.66	11.01	8.09	15.55	8.00
ζ_1 [-]	0.25	0.57	0.31	0.24	0.51	0.57	0.43	0.65	0.73
ζ_2 [-]	0.44	0.78	0.37	0.41	0.28	0.60	0.31	0.45	0.52
ζ_3 [-]	0.55	0.31	0.54	0.59	0.33	0.55	0.53	0.44	0.51
ζ_4 [-]	0.25	0.01	0.49	0.28	0.45	0.49	0.29	0.97	0.34
ζ_5 [-]	0.14	0.04	0.43	0.12	0.05	0.15	0.27	0.23	0.29

the standard error of the mean (SEM). The SEM is a measure of the variance of the parameter distribution. Generally, parameters that have little contribution to the prediction error, show large variances and a large SEM [Abbink, 2006; de Vlugt et al., 2006; Ljung, 1999]. The SEMs for the model parameters obtained for each task and direction are shown as black lines in Figs. 5.10-5.12. The results show a small SEM for almost all parameters, except for some parameters for the RT in lateral condition. For the vertical condition, the SEM is considerably higher for many parameters, signifying the limited reliability of the parameter identification in this direction [Venrooij et al., 2011c]. As mentioned earlier, the results for the vertical direction are exploratory only.

5.5 Model Validation

5.5.1 The global scope model

An evaluation of the model quality in the time domain can be obtained through the variance-accounted-for (VAF) [Abbink, 2006]. It

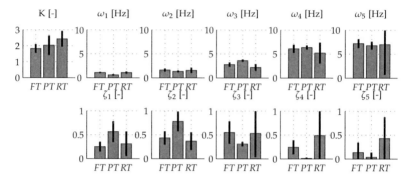

Figure 5.10: Model parameters obtained in lateral direction (measured with cyclic in roll direction). The black lines indicate the standard error of the mean (SEM).

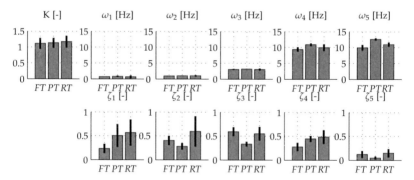

Figure 5.11: Model parameters obtained in longitudinal direction (measured with cyclic in pitch direction). The black lines indicate the standard error of the mean (SEM).

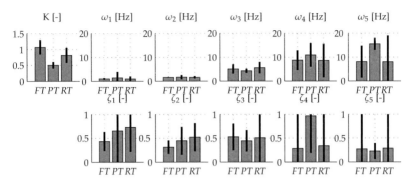

Figure 5.12: Model parameters obtained in vertical direction (measured with collective). The black lines indicate the standard error of the mean (SEM).

reflects how well the variance of a measured signal is approximated by its simulated counterpart. In this study the VAF is calculated for the B2P model, obtained by multiplying the B2FCL$^+$ model with the CD dynamics, Eq. 5.1. The model quality is determined by calculating the difference between the measured contribution of $M_{dist}(t)$ to the CD deflection, $\theta_{cd}^{M_{dist}}(t)$, and the modeled signal, $\bar{\theta}_{cd}^{M_{dist}}(t)$:

$$
VAF_{BDFT} = \left[1 - \frac{\sum_t \left[\theta_{cd}^{M_{dist}}(t) - \bar{\theta}_{cd}^{M_{dist}}(t) \right]^2}{\sum_t \left[\theta_{cd}^{M_{dist}}(t) \right]^2} \right] \cdot 100\% \qquad \text{5.12}
$$

The time signal $\bar{\theta}_{cd}^{M_{dist}}(t)$ can be obtained from simulation by providing the motion disturbance as input to the B2P model. The time signal $\theta_{cd}^{M_{dist}}(t)$ can be obtained through the frequency decomposition technique, described in Chapter 2.

The results shown in Fig. 5.9 and Table 5.3 were calculated using transfer function data that was averaged over all subjects. This will be referred to as the 'global scope model', indicating that the model was based on the most global set of data available. Note that this model only describes within-subject variability as all the between-subject variability has been lumped together through averaging over all subjects. To get an insight into how this 'lumping'

Table 5.4: VAF [%] for each subject in the lateral direction for the individual scope and global scope model.

	Subject 1		Subject 2		Subject 3	
	Indiv.	Global	Indiv.	Global	Indiv.	Global
FT	93.1	89.4	91.7	84.6	93.1	74.1
PT	86.4	87.0	82.9	78.0	78.5	64.7
RT	95.9	96.3	96.3	93.9	93.3	59.1

	Subject 4		Subject 5		Subject 6	
	Indiv.	Global	Indiv.	Global	Indiv.	Global
FT	94.9	92.8	93.9	94.5	85.3	74.3
PT	75.8	57.1	83.4	83.8	83.6	65.1
RT	96.1	88.4	96.4	96.0	94.5	73.4

	Subject 7		Subject 8		Subject 9	
	Indiv.	Global	Indiv.	Global	Indiv.	Global
FT	92.0	65.0	94.9	85.9	93.4	91.7
PT	85.8	53.4	86.1	77.0	84.5	84.7
RT	97.4	89.2	96.7	74.0	97.3	96.6

	Subject 10		Subject 11		Subject 12	
	Indiv.	Global	Indiv.	Global	Indiv.	Global
FT	93.4	73.0	94.7	91.7	88.0	47.6
PT	79.9	76.0	82.9	83.3	84.1	84.7
RT	93.5	76.9	96.7	95.6	95.1	77.6

degrades the model's quality, it is insightful to compare the performance of the global scope model, with that of 'individual scope models', i.e., models based on data obtained for a single subject.

By comparing the VAF values obtained by using the individual scope model and the global scope model, an indication is obtained of how the two models compare, or in other words, how *representative* the global scope model is when applied to a single subject. The results of this comparison for the lateral direction are shown in Table 5.4. The table lists the VAF value obtained using the individual scope and global scope model for each subject and each task. The results show that, as might be expected, the individual scope models provide a better performance than the global scope

Table 5.5: Average VAF [%] for each direction, obtained with the global scope model (standard deviations in parentheses).

	Lateral		Longitudinal		Vertical	
	Indiv.	Global	Indiv.	Global	Indiv.	Global
FT	92.4 (2.9)	80.4 (14.1)	73.8 (6.2)	60.2 (9.1)	38.4 (16.9)	20.7 (44.9)
PT	82.8 (3.2)	74.6 (11.6)	73.0 (7.6)	63.3 (16.9)	6.7 (14.7)	12.3 (14.8)
RT	95.8 (1.4)	84.8 (12.2)	78.0 (8.6)	50.7 (45.9)	32.9 (11.3)	13.7 (67.5)

model. For most subjects, however, the differences are only modest or even small (e.g., subjects 1, 5, 9 and 11), for some subjects the difference are larger (e.g., subjects 3 and 7). These differences are an indication of the magnitude and influence of between-subject variability. A commonly mentioned source of possible between-subject variability is body type, as it can be assumed that body size and mass influences BDFT dynamics [Mayo, 1989; Venrooij et al., 2013b]. However, no systematic relationships were found here between the performance of the global scope model and the somatotype (physical body type), which can be quantified using BMI [Maddan et al., 2008]. A similar observation was made in [Venrooij et al., 2013b], where it was observed that the BDFT data for the somatotypic groups showed only minor differences with respect to each other and the grand average BDFT.

By averaging the results in Table 5.4, the average VAF of the individual scope models and the global scope model is obtained, presented in Table 5.5. The table also shows the results after averaging the data in longitudinal and vertical direction. For the lateral and longitudinal directions the VAF values obtained for the individual scope model are high, and the VAF values obtained using the global scope model are slightly lower. It should be noted that the large standard deviation obtained for the global scope model in the longitudinal direction indicates a large spread between subjects, and hence, that the global scope model for some subjects did not perform well. However, it is possible to conclude that for these two directions the global scope model provides, in general, an adequate description of the BDFT dynamics when applied to individual subjects.

Table 5.6: VAF [%] for the test pool subjects in the lateral direction for the partial scope model.

	Subject 1 Partial	Subject 4 Partial	Subject 7 Partial	Subject 10 Partial
FT	89.6	90.1	58.5	66.8
PT	87.5	58.1	50.1	75.4
RT	96.4	88.1	89.1	76.4

The results obtained for the vertical direction require some more elaboration. For this direction the VAF values are low for both the individual scope and the global scope models. As already mentioned before, the data in the vertical direction show low coherences, especially for the PT (see also [Venrooij et al., 2013b]). A low coherence indicates a low signal-to-noise ratio (SNR) in the data [Venrooij et al., 2011c], i.e., that a relatively large amount of noise is present in the signal. This noise manifests itself in the frequency domain through jitter in both magnitude and phase parts of the Bode plot, as can be seen for the vertical direction in Figs. 5.3 and 5.4. Data with higher coherences, i.e., relatively less noise, show much smoother graphs, like those obtained for the lateral direction. By fitting a model on noisy data, a 'smoothed' noise-free model approximation of the dynamics is obtained. The model does not (and should not) follow the noise induced jitter of the measured data, but instead provides a smooth transfer function, see Fig. 5.9. However, when calculating the VAF one is comparing a noisy time-signal with noise-free model approximation, and hence, a low VAF does not necessarily imply a poor model. The presence of noise alone degrades the VAF and for low SNR signals this influence can be substantial. There are some possible ways of dealing with this issue, but they were not pursued here. In this chapter, the VAF is presented as computed from the noisy measured data, while hypothesizing the low values observed for the vertical direction are caused mostly by a low SNR and not by a flawed model. This hypothesis is supported by the frequency domain fit shown in Fig. 5.9, where it can be observed that the models seem to be the best possible smooth approximation of the measured dynamics.

Table 5.7: Average VAF [%] for the test pool subjects, obtained with the global scope model (standard deviations in parentheses).

	Lateral		Longitudinal		Vertical	
---	Global	Partial	Global	Partial	Global	Partial
FT	80.1 (13.2)	76.3 (16.1)	62.7 (6.2)	59.0 (5.1)	28.7 (39.1)	27.8 (37.2)
PT	68.3 (15.9)	67.8 (16.9)	59.4 (25.3)	58.8 (25.6)	24.8 (15.3)	24.1 (15.1)
RT	87.7 (8.0)	87.5 (8.3)	73.5 (12.3)	73.8 (10.7)	44.1 (18.9)	49.8 (14.6)

5.5.2   The partial scope model: based on subject subgroup

The results in the previous section give an indication of how representative the global scope model, as presented in Table 5.3, is for individual subjects. To further ensure that the model is not just an adequate fit but also has some predictive capabilities, also a 'partial scope model' was constructed. This model was constructed using exactly the same techniques as described for the global scope model, but now only including the data of 8 of the 12 subjects. First, four subjects were selected as test pool subjects. In this case, each third subject, starting with the first, i.e., subject 1, 4, 7 and 10. As the subject were numbered based on random scheduling, this is a quasi-random selection. A partial scope model was constructed using only the remaining subjects, and this model was applied to the test pool. The resulting VAF values show how well this partial scope model performs on data it has never 'seen'.

The VAF obtained using the partial scope model on the test pool is shown in Table 5.6. By comparing these numbers with those in Table 5.4, it can be concluded that the partial scope model shows approximately the same performance as the global scope model. These results show the 'predictive capabilities' of the partial scope model, when applied to 'new' subjects. It is fair to assume a similar performance for the global scope model on other subjects than used in this study.

Just as was done for the global scope model, one can average the results for each disturbance direction. The results of this procedure for the partial scope model are listed in Table 5.7. It should be noted that the average is now taken over the four test pool subjects only, therefore the results for the global scope model differ from those in

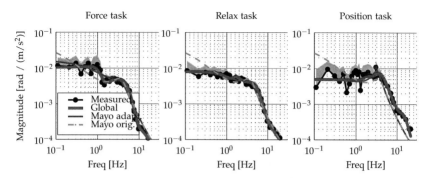

Figure 5.13: Comparing the original Mayo model [Mayo, 1989], the adapted Mayo model [Venrooij et al., 2013b] and the global scope model proposed in the current chapter with the measured magnitude data for the vertical direction.

Table 5.5. The results confirm the observation that the partial scope model performance is satisfactory and about equal to the global scope model for all directions.

5.5.3   A comparison with other BDFT models

For a few cases the performance of the currently proposed global scope BDFT model can be compared with that other BDFT models. In the frequency domain, the model can be compared to the models proposed by Mayo [1989] and the adaptation proposed in [Venrooij et al., 2013b]. The first will be referred to as the 'original' Mayo model, the latter as the 'adapted' Mayo model. Note that the adapted Mayo model was based on the same experimental data as

Table 5.8: Average VAF [%] for lateral direction, obtained with the physical and global scope model (standard deviations in parentheses).

	Lateral		
	Indiv.	Global	Physical (Chptr. 4)
FT	92.4 (2.9)	80.4 (14.1)	82.75 (6.3)
PT	82.8 (3.2)	74.6 (11.6)	88.48 (3.8)
RT	95.8 (1.4)	84.8 (12.2)	87.29 (3.7)

used in the current chapter, but without making any changes to Mayo's model structure, which is [Mayo, 1989]:

$$H_{mayo,orig}(s) = \frac{b_1 s + b_2}{s^2 + a_1 s + a_2}$$

5.13

Both the original and adapted variation of the Mayo model were only defined for the vertical direction, so the comparison is limited hereto. Recall that this is the direction where the data are of poor quality.

Fig. 5.13 shows the comparison between the original Mayo model [Mayo, 1989], the adapted Mayo model [Venrooij et al., 2013b] and the global scope model for the vertical direction. Only magnitude information is shown here. Note that the measured data and the global scope model were already shown in the upper right of Fig. 5.9. Clearly, the global scope model shows a superior fit to the measured data. The original Mayo model provides a rather poor fit across the different control tasks. The adapted Mayo model shows a marked improvement but is not capable of following the measured dynamics for the high frequencies (> 4 Hz) for the PT. From Fig. 5.13, it can be concluded that the model proposed in the current chapter provides the best approximation. The VAF values of the models could be used to confirm this, but they were not calculated here due to the limited validity of the VAF values observed in this particular direction.

In Chapter 4 (and [Venrooij et al., 2014a]) a *physical* BDFT model was proposed. This model aims to represent BDFT dynamics using a physical approach, representing limb masses, muscle dynamics and reflexes. This model was only developed for the lateral direction, so the comparison is limited to this direction only. Table 5.8 shows the average VAF values obtained by averaging the results of individual physical models, taken from Chapter 4. For convenience, the VAF data from Table 5.5 are repeated here. It should be noted that the physical model was constructed for measurements with a side-stick instead of a cyclic. While keeping that in mind, it is still insightful to compare the VAF values to get a feel for how the two models compare. The VAF values presented for the physical model were obtained by constructing an individual scope model for each subject [Venrooij et al., 2014a], so these results should be compared with the average VAF of the individual scope models from the current study. The results for the current model are slightly better for the FT and RT, but the differences are modest. This shows that a mathematical model, although lacking a physical basis, can provide excellent and reliable models. An even more important fact to observe is the effort involved in estimating the parameters of the two models. The physical model consists of 18 parameters, making it comparable in complexity to the current mathematical model of 16 parameters. Finding the proper values of the *physical* parameters, however, requires a considerable number of assumptions and moreover a two-stage parameter estimation approach, in which first an admittance model needs to be constructed (for details the reader is referred to Chapter 4). Estimating the parameters of the physical model is a considerably more complex and involved procedure than that of the mathematical model proposed in the current chapter.

Does this mean the physical model in Chapter 4 is now obsolete? No, definitely not. As argued earlier, physical models provide a different kind of information than black box or mathematical models do. The insights provided by the physical model regarding the physical principles that govern BDFT can in no way be paralleled by what the mathematical model provides. The physical model remains a useful, sometimes even necessary tool in BDFT investigations. However, if one is interested in BDFT modeling without a strong interest in the physical details of it – as many researchers and engineers in practice are – the mathematical model provides a simpler, more usable alternative.

5.6 Conclusions and recommendations

In the current chapter a mathematical BDFT model was proposed. This model aims to fill the gap between currently existing simple but therefore often limited black box models and the versatile but more complex physical models. Through asymptote modeling a structural method was obtained to construct a high fidelity model, with limited complexity, allowing for a reliable parameter estimation and a straightforward implementation. Suitable parameter sets for the model structure were proposed and by using these the model can be directly implemented in many typical rotorcraft BDFT studies.

The model performance was reported in both the frequency and the time domain. The 'global scope model', obtained by averaging the results of all subjects together, performs well, although 'individual scope models' are slightly superior. By evaluating a 'partial scope model', based on only a part of the subject pool, further confidence was gained that the model is applicable to other subjects than those used in this study.

Furthermore, the model's performance was compared to two black box models in the frequency domain. The results show that the currently proposed model is superior. Finally, the model was compared to the physical model proposed in Chapter 4 in the time domain. In this case the performance proved to be comparable, but it is noted that the parameter estimation of the mathematical model is considerably easier. The physical model is still an important tool in BDFT investigations, as it provides a physical insight that the current mathematical model cannot provide. However, in many cases engineers, designers and researchers, are more interested in modeling the overall effects of BDFT, instead of scrutinizing their underlying physical causes, and there the currently proposed model may be a practical, more usable alternative.

An important limitation of the model is its dependency on control device dynamics (note that this limitation does not exist for a physical model). The performance of the model with different settings of the control device needs to be addressed in future studies. In the current study, linear dynamics without friction were used for the control devices. This may not be representative for actual cases, and the impact of this simplification needs to be investigated. Also, judged from the obtained coherences, the data measured in the vertical direction were of poor quality. This reduces the reliability of the model parameters in this direction considerably. A follow-up study focusing on the vertical direction may improve the model's quality. The approach proposed to obtain the model structure through asymptote modeling can be tested and improved by application to other problems in other fields.

Part III

Mitigating
biodynamic feedthrough

CHAPTER 6

Biodynamic feedthrough mitigation techniques

In this chapter the range of possible BDFT mitigation techniques is identified and evaluated. Two mitigation techniques are regarded to be most promising: passive support/restraining systems and model-based BDFT cancellation. The potential of both techniques is investigated. First the potential effects of an armrest on BDFT dynamics are discussed. Then model-based signal cancellation is evaluated in simulation, using optimal signal cancellation. The results of the latter analysis show that signal cancellation is a promising mitigation method for BDFT problems, but only if the model can be adapted to both subject and task.

The contents of this chapter are based on:

Paper title	A Review of Biodynamic Feedthrough Mitigation Techniques
Authors	Joost Venrooij, Max Mulder, Marinus M. van Paassen, Mark Mulder, and David A. Abbink
Published at	11th IFAC/IFIP/IFORS/IEA Symposium on Analysis, Design, and Evaluation of Human-Machine Systems, Valenciennes, France

- and -

Paper title	Cancelling Biodynamic Feedthrough Requires a Subject and Task Dependent Approach
Authors	Joost Venrooij, Mark Mulder, Marinus M. van Paassen, David A. Abbink, Heinrich H. Bülthoff, and Max Mulder
Published at	2011 IEEE International Conference on Systems, Man and Cybernetics, October 9-12, 2011, Anchorage, USA

6.1   Introduction

Previous chapters have provided an insight in some of the complexities of biodynamic feedthrough (BDFT). It was shown that many different factors influence the BDFT dynamics, such as the control device dynamics, disturbance direction and disturbance frequency. Also, we have seen that BDFT is strongly influenced by the highly variable human body dynamics, an influence we are only beginning to understand. The limited understanding of these different aspects of BDFT becomes most apparent in BDFT mitigation, i.e., when trying to solve the BDFT problem. This chapter addresses the possible techniques for BDFT mitigation. First, it aims to answer the question which possible mitigation approaches we have at our disposal. By evaluating the benefits and disadvantages of each and testing these against a set of requirements, the promising approaches are then selected for further investigation. The following four requirements (which will be detailed in the chapter) were used:

- the approach should relieve the human operator from its role as BDFT mitigator,

- the approach should be generally applicable across different vehicles,

- the approach should have a manageable complexity and require limited modifications to the vehicle, and

- the approach should not purely rely on the separation assumption.

The separation assumption can be defined as follows [Venrooij et al., 2010b]:

Separation assumption

A clear bandwidth separation exists between voluntary, cognitive control inputs and involuntary, BDFT induced control inputs.

After selection of promising mitigation methods, the potential of these methods will be addressed using literature, simulation and the BDFT framework that was developed in Chapter 3.

6.2   Potential mitigation approaches

To discover the potential BDFT mitigation approaches, use can be made of the BDFT system model, discussed in detail in Chapter 3. By addressing each element in the BDFT system model, the possible approaches can be discovered and categorized. In total, seven *solution types* can be identified, each allowing for one or several *solution approaches* [Venrooij et al., 2010b]. The seven types are indicated with numbered stars in the BDFT system model presented in Figure 6.1. Recall that BDFT occurs when accelerations, i.e., motion disturbance signal $M_{dist}(t)$, feed through the human body, causing involuntary forces applied to the control device and, in turn, involuntary control device deflections, which enter the CE.

Three possible solution types act at one of the elements through which the motion disturbance passes before it enters the CE: the PLF (#1), the NMS (#3), and the CD (#5). BDFT mitigation can also be applied between these elements, yielding three additional solution types: between the PLF and the NMS, i.e., at H_{PLFHO} (#2), between the NMS and the CD, i.e., at H_{HOCD} (#4) and between the CD and the CE (#6). These types rely on reducing the feedthrough of acceleration (or the effects thereof) between the different physical system elements. Finally, there is model-based cancellation (#7), which uses a model to determine a canceling signal based on the vehicle accelerations. This solution type is indicated by dotted lines in Figure 6.1. It differs from the other solution types as it does not reduce BDFT directly at or between the involved elements, but by adding an additional system to calculate a canceling signal. In the following, each of the seven solution types are discussed, including their most important benefits and disadvantages. The results are summarized in Table 6.1 on page 228.

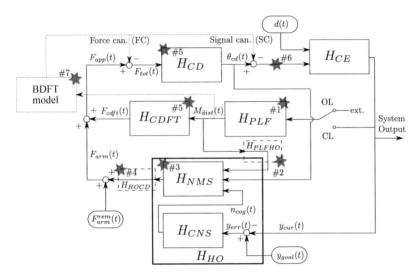

Figure 6.1: The biodynamic feedthrough system model. A human operator (HO) controls a controlled element (CE) using a control device (CD). Motion disturbances $M_{dist}(t)$ are coming from the platform (PLF). The feedthrough of $M_{dist}(t)$ to involuntary applied forces $F_{arm}(t)$ and involuntary control device deflections $\theta_{cd}(t)$ is called biodynamic feedthrough (BDFT). The feedthrough of $M_{dist}(t)$ to inertia forces $F_{cdft}(t)$ is called control device feedthrough (CDFT). $F_{app}(t)$ is the sum of the forces applied to the control device by the HO. The HO consists of a central nervous system (CNS) and a neuromuscular system (NMS). The connection between the HO and the environment is governed by two 'interfaces', H_{PLFHO} and H_{HOCD}. The CE and PLF can form an open-loop (OL) or closed-loop (CL) system. The focus of the current chapter: potential BDFT mitigation solution types are discussed, which are indicated here with numbered stars.

6.2.1 Minimizing platform accelerations (#1)

Platform accelerations are the source of all BDFT effects. Minimizing these accelerations is arguably the most straightforward and practical way of mitigating any adverse acceleration induced effects. By preventing the accelerations that (could) cause BDFT from occurring in the first place, BDFT is effectively removed as a potential problem. However, in practice, reducing the PLF acceleration often involves sacrificing the system's agility and responsiveness. For some vehicles it may be acceptable to trade these in return for increased safety (e.g., for electric-powered wheelchairs or hydraulic platforms). However, for the majority of vehicles, responsiveness may be of vital importance (e.g., fighter jets). The applicability of minimizing platform accelerations as a solution approach to BDFT is therefore mainly limited to vehicles whose responsiveness or agility is not a primary interest, and even for those vehicles it may be beneficial to deal with BDFT using one of the other mitigation techniques instead.

6.2.2 PLF-HO interface design (#2)

By carefully designing the interface between the platform and the human operator (H_{PLFHO}), the feedthrough of platform accelerations into the body of the human operator can be reduced. Common examples are passive support/restraining systems such as seat damping and seat belts. Seat damping improves ride comfort and mitigates BDFT effects by reducing the propagation of accelerations through the seat. A restraining system such as seat belts (partially) immobilizes the torso and hence reduces the propagation of accelerations into involuntary upper body motion. These passive support/restraining systems are a simple and cost effective way of reducing BDFT and are present in many common vehicles. It should be noted that these systems are usually not installed for BDFT mitigation alone, but mainly for other reasons such as safety and comfort. The fact that these systems may also reduce the propagation of vehicle accelerations into involuntary control inputs can be regarded as an 'added bonus'. It should be noted, however, that

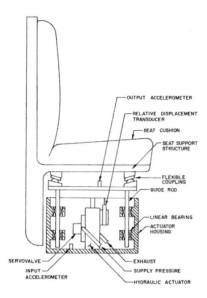

OUTPUT ACCELEROMETER
RELATIVE DISPLACEMENT TRANSDUCER
SEAT CUSHION
SEAT SUPPORT STRUCTURE
FLEXIBLE COUPLING
GUIDE ROD
LINEAR BEARING
ACTUATOR HOUSING
SERVOVALVE
EXHAUST
INPUT ACCELEROMETER
SUPPLY PRESSURE
HYDRAULIC ACTUATOR

Figure 6.2: Major functional components of the electro-hydraulic pilot seat as proposed in [Schubert et al., 1970], used to isolate the human operator from vehicle accelerations.

BDFT still occurs in vehicles equipped with these systems [Raney et al., 2001], hence, these passive system are not always sufficient to completely remove all BDFT effects.

A more rigorous way of preventing vehicle accelerations from entering the human body is to actively isolate the HO from the PLF accelerations. A system based on this approach was developed and investigated by Schubert et al. [1970], see Fig. 6.2. The Active Vibration Isolation System (AVIS) actively compensates for platform accelerations in the vertical direction, such that the human operator is isolated from the accelerations. In principle, this approach allows for effective removal of BDFT effects, but its implementation comes with many practical difficulties. Dimasi et al. [1972] used the AVIS to test various body isolation configurations consisting of combinations of torso, hand and foot isolation. The results suggested that isolation yields an improvement in ride comfort, as was hypothesized. However, evaluation of tracking performance showed significant improvement only if *all* the elements with which the HO interacts, i.e., the seat, the displays and the controls, were isolated. It was observed that the feedthrough of acceleration depends on the relative motion between the HO and these elements, rather than the HO's inertial motion alone.

There are several drawbacks when isolating the HO from vehicle accelerations in an attempt to minimize BDFT. The first is the complexity (and thus costs) of a mechanical system that is required to compensate for fast and often highly stochastic platform accelerations. Secondly, and more importantly, platform accelerations also form an important source of *information* on the state of the vehicle through motion cues. Isolating the human operator from platform accelerations removes these essential cues, possibly degrading the operator's situational awareness and deteriorating control performance in complex situations.

In [Humphreys et al., 2011] an alternative solution for the occurrence of BDFT in backhoes (a type of excavator) is proposed. The approach is similar to the AVIS proposed in [Schubert et al., 1970] in that it reduces that exposure of the operator to vehicle vibrations, but without requiring an additional complex and expensive compensation system. The solution proposed uses the backhoe arm for a dual task: for performing excavation operations and for cabin vibration reduction[a]. For a successful vibration reduction while maintaining an adequate level of responsiveness a trade-off between working performance and cabin vibration reduction needs to be established. First results suggest that a significant reductions in cabin motion can be obtained with minimal tracking performance degradation. Notwithstanding these promising results, it should be

---

[a]Note that the solution approach proposed in [Humphreys et al., 2011] may also be classified as 'minimizing PLF accelerations', as the approach aims at minimizing the accelerations the human operator is subjected to. However, as the proposed approach does not rely on reducing the acceleration of the vehicle directly, but uses a controller to reduce the amount of vibrations the HO is exposed to, it is categorized here as a PLF-HO interface design solution.

noted that this approach is specific for excavator(-like) vehicles and not applicable to vehicles in general.

6.2.3   Neuromuscular adaptation (#3)

Probably the most common mitigation technique for BDFT problems, used on a daily basis, is provided by the human operator him-/herself. In many BDFT situations, across different vehicles, occupants react by attaining a setting of the neuromuscular system which minimizes the impact of accelerations on control performance. For example, helicopter pilots are trained to hold the control devices loosely (often with only two fingers) while keeping the arm relaxed and supported on their leg or knee. This is not only a matter of comfort, but also an effective way to minimize certain BDFT effects. In fact, many Pilot-Assisted-Oscillations (PAOs) in rotorcraft are remedied through 'procedural mitigations', which are "recommendations to modify the pilot's behavioral response to the PAO by suggesting removal of the pilot's hand from the cyclic stick or relaxing the pilot's grip on the cyclic stick and reducing maneuver severity to disengage the pilot from the PAO" [Walden, 2007]. An example is the procedural mitigation implemented in the V22B-Osprey, recommending the pilot to relax the grip on the cyclic to mitigate the so-called 1.4 Hz high focal roll mode oscillation [Walden, 2007]. Such an instruction to 'relax' the neuromuscular system is similar to the relax task (RT) instruction used in the experiments throughout this thesis. The RT elicits a neuromuscular setting that reduces the feedthrough of accelerations especially at higher frequencies (Fig. 2.8 on page 54). The same figure also shows, however, that in order to mitigate BDFT for lower frequencies it would be beneficial to attain a stiffer neuromuscular setting. In general, there is not one setting of the neuromuscular system which reduces BDFT for both low and high frequencies.

We can safely state that neuromuscular adaptation is currently the best (and possibly the only) truly adaptive BDFT mitigation technique available. The capability of humans to easily, quickly and accurately adapt their body dynamics is almost impossible to match by even the most modern technologies. In fact, neuromuscular adaptation may be the main reason why we, in our daily lives, rarely experience BDFT as a severe problem. Often, neuromuscular adaptation prevents bad from turning into worse and helps to reduce BDFT events to minor annoyances. Without it, walking while holding a cup of hot coffee would be disastrous. Only by adapting the neuromuscular settings one prevents spilling the coffee. Similar adaptive neuromuscular mechanisms are employed when stabilizing a helicopter in strong turbulence. Neuromuscular adaptation is an important mechanism when performing manual control tasks in motion environments.

However, although neuromuscular adaptation is a powerful mitigation technique, relying on the human operator to deal with all BDFT effects is not a real solution either. Firstly, the fact is that BDFT does exist, even for well trained human operators, implies that neuromuscular adaptation alone is not enough. Secondly, neuromuscular adaptation puts constraints on the operator's skill and behavior, and may require sacrificing other aspects of the control performance. The optimal neuromuscular setting is dependent on many other factors, such as the required control precision and speed. As a result, a 'relaxed' setting of the neuromuscular system is not always optimal in terms of the requirements posed by the situation at hand, just as a 'stiff' setting is not. It can be said that attaining a particular neuromuscular setting for the sake of BDFT mitigation may involve sacrifices, e.g., a reduction in control bandwidth, control speed or accuracy.

6.2.4   HO-CD interface design (#4)

In the design of the interface between the human operator and the control device (H_{HOCD}) similar BDFT mitigation measures can be implemented as discussed for the PLF-HO interface. These measures rely on supporting or immobilizing the limb that is in contact with the control device. A good example of this solution type is an armrest, a common device installed in many vehicles, e.g., aircraft. Armrests increase comfort and reduce fatigue, but they are also effective in stabilizing the arm when subjected to motion disturbances. Chapter 7 will show the results of a study where the effectiveness of an armrest to mitigate BDFT was studied.

An alternative way to prevent the propagation of accelerations through the control limb, the limb can also be restrained (tied down), just as is often done with the torso by means of seat belts. Applications of this approach are, however, very limited as such restraints also limit the *voluntary* actions of the operator (e.g., reach out to press a button). In a study in which the effects of a restraining harness on tracking performance were measured during vibrations [Lovesey, 1971a,b] it was found that the addition of the harness increased the transmission of the vibrations through the body, generally deteriorating control performance.

Just as the support/restraining approaches discussed in Section 6.2.2, the measures discussed here are simple and cost effective, but often not sufficient to completely remove all BDFT effects (see also Chapter 7).

6.2.5 Control device design (#5)

For the control device, several options to mitigate BDFT are imaginable. For example, research has suggested that the control gain, i.e., the overall sensitivity of the control device, can be optimized to increase tracking performance in vibration conditions [Lewis and Griffin, 1977]. It was found that there is a considerable interaction between the effects of motion disturbances and the control gain, implying that the optimal gain varies with the characteristics of the motion disturbance. In general, the optimal control gain for minimizing tracking error under a vibration condition is likely to be lower than that for a static condition [Lewis and Griffin, 1977]. This dependency makes adjustments of the control gain an impractical BDFT mitigation approach, as continuously adapting the control gain is likely to have detrimental effects on voluntary control performance.

An alternative option is to reduce the control device responsiveness to only a part of the frequency contents of the input forces (e.g., using a notch filter). In this way the level of involuntary control input that propagates through the CD and enters the CE can be reduced for particular frequencies. An important assumption when applying this method is the separation assumption, which asserts that a clear bandwidth separation exists between voluntary, cognitive control inputs and involuntary, BDFT induced, control inputs [Venrooij et al., 2010b].

An example of a study relying on the separation assumption is [Velger et al., 1984], where it was assumed that cognitive control activity is limited to 1 Hz and vibration induced control activity only occurs at higher frequencies. There are several arguments against this assumption. First of all, studies have shown that reflexive activity may play an important role in the operator's control behavior, yielding relevant dynamics at frequencies also above 1 Hz (e.g., [Mugge et al., 2009]). Furthermore, several studies measured significant BDFT effects at frequencies lower than 1 Hz [Sövényi and Gillespie, 2007; Venrooij et al., 2009]. The low frequency BDFT effects are sometimes assumed to be irrelevant by arguing they can be cognitively corrected for by the human operator. Even if an operator is capable of doing so, this imposes a mental load on the operator which adds to the effort required to obtain a certain control performance. In some cases the corrective control inputs may prevent a degradation in performance, but it always comes at the expense of effort. Simply ignoring low frequency BDFT effects is therefore not always justified.

It is likely that solution approaches which rely on the separation assumption either cannot fully eliminate (low frequency) BDFT effects or also partly suppress the voluntary (high-frequency) control signal.

An alternative adaptation to the control device that can be made is its dynamics. A stiffer control device will deflect less in response to an involuntary force. Although this reduces BDFT, it also negatively affects controllability as a stiffer control device does not distinguish between voluntary and involuntary control inputs. As mentioned by McLeod and Griffin [1989], the optimal dynamical setting is a compromise between BDFT resistance and controllability. It can be concluded that optimizing the control device dynamics to minimize BDFT effects often has negative effects on controllability [McLeod and Griffin, 1989; Torle, 1965].

The third and final approach discussed here regarding the control device is ensuring minimal alignment between the manual control axes and the axes of PLF motion. The strongest decrement in control performance occurs for motion disturbances in the same direction as the sensitive axis of the control and display [Lewis and Griffin, 1977]. A steering wheel is immune to accelerations in the longitudinal direction, as it has no degree of freedom aligned with this axis. This effectively eliminates the occurrence of BDFT in this direction. This is not the case in, e.g., a side-stick, rendering that type of control device more susceptible to BDFT. By carefully selecting the control device and positioning its axes, BDFT could in some cases be suppressed. However, in practice, the selection of the control device is based on many other factors than its susceptibility to BDFT. In most cases, the alignment of control and the motion axes cannot be prevented. In fact, aligning the axes of control with the axes of motion is often the most intuitive, and thus the preferable design from a human-machine interface perspective.

In some cases the implementation of novel and more intuitive control devices is complicated by their susceptibility to BDFT. For example, the BDFT study presented in [Humphreys et al., 2011] is motivated by an attempt to replace the conventional backhoe user interface consisting of two separate 2 degree-of-freedom (DOF) joysticks, with a single 6-DOF haptic input device. Although studies showed that the 6-DOF input device provides more intuitive operation, it was also shown that they are more susceptible to BDFT effects, which negates the control performance benefit. The study presented in [Humphreys et al., 2011] was performed to solve BDFT problems induced by using an otherwise superior control input device. A better understanding of how to deal with BDFT may allow these and other vehicles to benefit from more intuitive, lighter or responsive controls.

6.2.6 Signal filtering (#6)

Between the control device and control element, the control input signal can be filtered. This method is used in practice, for example in helicopters where notch filters are implemented to decouple the pilot interaction with structural modes. Examples of naval rotor-craft where notch filters were considered or implemented to remove Pilot-Assisted-Oscillations (PAOs) are the CH-46D Sea Knight, SH-60B Seahawk, CH-53E Super Stallion, V-22A/B Osprey, and the AH-1Z Viper [Walden, 2007]. These filters remove particular frequencies from the inputs. Note that such a filter does not distinguish between BDFT related inputs and voluntary inputs. Signal filtering may have the advantage that its implementation is relatively easy and thus cost effective. A disadvantage is that this approach relies on the separation assumption and that filters may remove voluntary control inputs. In addition, they may introduce additional delays in the control loop.

A slightly different filtering approach was proposed by Velger et al. [1984], who implemented an adaptive filtering technique to suppress BDFT effects. In this approach, a least mean square (LMS) adaptive filter is used that adapts based on measurements of the platform accelerations. This 'adaptive input filtering' approach offers a somewhat more versatile alternative to the 'static input filtering', but in closed-loop cases the success of this method still relies on the separation assumption. An additional possible issue with any adaptive approach is certification, as it needs to be shown that the functionality is guaranteed for the complete adaptive range.

In [Velger et al., 1988] the adaptive filter was evaluated in some closed-loop experiments. In line with the separation assumption, the PLF motions were high-pass filtered before being used as input for the adaptive filter, leaving low frequency BDFT rejection to be handled by the operator. Some additional tests were performed without the high-pass filter present and it was reported that subjects found that the adaptive filter interfered with their commands at lower frequencies.

From the above it can be concluded that signal filtering has shown to be effective in several studies. However, an important disadvantage of any filtering approach is that it does not affect involuntary BDFT inputs only. A filter that attenuates certain frequencies from a signal does not distinguish between voluntary and involuntary control inputs. As voluntary and involuntary control inputs can (and usually do) overlap in frequency this can lead to suppressing voluntary control inputs.

6.2.7   Model-based cancellation (#7)

Several studies have investigated model-based BDFT cancellation (e.g., [Gillespie et al., 1999; Sirouspour and Salcudean, 2003; Sövényi and Gillespie, 2007]). This approach uses an electronically generated canceling signal which is injected in the human-vehicle system to cancel BDFT effects. This approach differs from signal filtering (#6) as it relies on a model to calculate the involuntary BDFT part in the control signal, instead of filtering this part from the signal directly[b]. By adding the cancellation signal to the actual signal, which contains both voluntary and the involuntary parts, BDFT is canceled. An advantage of this approach is that it does not rely on the separation assumption. With an accurate model, involuntary control inputs can be successfully canceled without removing any of the voluntary control inputs. The control device dynamics and vehicle dynamics remain unaltered, just as the cockpit layout. Except for the installation of an Inertial Measurement Unit (IMU) to obtain a measurement of the PLF accelerations (which may already be present for other purposes) no other weight needs to be added to the vehicle.

A disadvantage of the model-based cancellation method is that its success relies largely on the accuracy of the model, making an accurate model a necessity. As BDFT is a complex process, capturing its dynamics accurately in a model is challenging. Different types of BDFT models have been developed over the past decades, e.g., [Jex and Magdaleno, 1978; Mayo, 1989; Sirouspour and Salcudean, 2003; Sövényi and Gillespie, 2007], but only few of them have been actually tested in model-based cancellation.

---

[b]Note that the implementation of the adaptive filter in [Velger et al., 1984] and [Velger et al., 1988] is such that it could also be categorized as model-based cancellation, where the combination of the high-pass filter and adaptive filter constitute the model.

The cancellation itself can be achieved through two distinct mechanisms, both are indicated in Figure 6.1: force cancellation (FC) and signal cancellation (SC).

Force cancellation

This approach is also known as force reflection and is based on mechanically inserting a canceling force or torque at the control device that counters the BDFT part of the force or torque applied by the control limb of the human operator. Note that force cancellation can only be applied to control devices that are movable, i.e., are not isometric (stiff). One can cancel the involuntary force applied to a stiff control device, but that should be classified as signal cancellation rather than force cancellation, as one is subtracting an involuntary signal instead of reflecting an involuntary force. The method of force cancellation is proposed and tested in [Gillespie et al., 1999; Repperger, 1995; Sövényi, 2005; Sövényi and Gillespie, 2007]. These studies show that this approach is promising, but that applying canceling forces also changes the 'feel' of the control device. It is largely unknown whether and how the human operator will adapt to these haptic changes and how this adaptation, in turn, will influence the occurrence of BDFT.

One particularly elegant aspect of the force cancellation approach is that it offers a way of involving the human operator in the cancellation process. Through the forces on the control device the operator is informed about the controller's activity. As the operator also exerts control over the vehicle through forces applied to this control device, the control authority is 'shared' between operator and controller. In other words, the model-based force cancellation approach adheres to the shared control paradigm [Griffiths and Gillespie, 2004]. In many other controller-based solutions the operator is excluded from the cancellation control loop, reducing the operator's awareness of the activity of the controller.

Signal cancellation

An alternative canceling approach is to subtract the modeled BDFT control input from the total control input, before it enters the controlled element. This method was used in [Sirouspour and Salcudean, 2003] and is referred to here as the signal cancellation (SC) method. It fundamentally differs from the force cancellation method in the sense that it does not cancel the effect of BDFT at the control device, but *after* the control device. Therefore, this approach does not alter the 'feel' of the control device. Another, more practical advantage of this approach is that it does not require an 'active' control device, i.e., a control device that is capable of generating forces. The cost, complexity, and additional weight of such a control device may make it impossible or very costly to implement a force cancellation approach, making signal cancellation preferable. A disadvantage of this method is that it excludes the human operator from the corrective control loop. An experiment in which model-based signal cancellation was applied as mitigation method is presented in Chapter 8.

6.3   Selection of promising approaches

Table 6.1 summarizes the results from the previous section. Currently, many vehicles already offer some of the approaches mentioned. Most used are the passive measures that restrain and support body parts (#2.1, #2.2 and #4.1). Note that these measures by themselves are not sufficient to remove BDFT effects completely. For example, in a study by Raney et al. [2001] BDFT occurred while the pilots were strapped-in tightly in (cushioned) aircraft seats equipped with armrests. So, even with these measures present BDFT occurs. To handle this, the human is often required to adapt his or her neuromuscular settings (#3). Although this process is fast, reliable and does not required adaptations to the vehicle, part of the control performance (e.g., bandwidth) is likely to be sacrificed. Furthermore, the occurrence of BDFT for well-trained pilots has demonstrated that neuromuscular adaptation cannot be relied on to remove all BDFT effects. It would be desirable to relieve

Table 6.1: Summary of biodynamic feedthrough solution approaches.

ID	§	Type	Principle	Example / Literature	Major advantage	Major disadvantage
#1	§6.2.1	Minimize PLF accelerations	Limit $M_{dist}(t)$; prevent BDFT	Acceleration limiter wheelchair	Direct BDFT prevention	Sacrifice responsiveness
#2	§6.2.2	PLF-HO design	Reduce propagation of $M_{dist}(t)$			
#2.1	§6.2.2	- damping	Immobilize body parts	Seat cushioning	Simple; passive; low-cost	Not always sufficient
#2.2	§6.2.2	- restraints	Immobilize body parts	Seat belts [Schubert et al., 1970]	Simple; passive; low-cost	Not always sufficient
#2.3	§6.2.2	- isolation	Isolate HO from PLF		Effective BDFT removal	Complex, expensive
#3	§6.2.3	Neuromuscular adaptation	Adapt body dynamics	Procedural mitigation in rotorcraft	Adaptive, fast, no vehicle modification	Sacrifice performance / effort
#4	§6.2.4	HO-CD design	Reduce invol. limb motion	Armrest	Simple; passive; low-cost	Not sufficient
#4.1	§6.2.4	- supports				
#4.2	§6.2.4	- restraints	Immobilize body parts	[Lovesey 1971b]	Simple; passive; low-cost	Limits voluntary movements
#5	§6.2.5	CD design				
#5.1	§6.2.5	- CD responsiveness	Decrease responsiveness	[Walden, 2007]	Simple; passive; low-cost	Separation assumption
#5.2	§6.2.5	- CD dynamics	Increase inertia or stiffness	[Walden, 2007]	Simple; passive; low-cost	Negative effects on controllability
#5.3	§6.2.5	- axis design	Prevent alignment control axis with acceleration axes	Steering wheel	Axis immunity to BDFT	May be impractical / unintuitive
#6	§6.2.6	Signal filtering				
#6.1	§6.2.6	- static filtering	Filter θ_{cd} with static filter	Walden [2007]	Simple; passive; low-cost	Separation assumption
#6.2	§6.2.6	- adaptive filtering	Filter θ_{cd} with adaptive filter	[Velger et al., 1988]	Adaptive solution	Certification
#7	§6.2.7	Model-based cancellation				
#7.1	§6.2.7	- force cancellation	Oppose involuntary force	[Sövényi and Gillespie, 2007]	Operator in loop (shared control)	Active CD required
#7.2	§6.2.7	- signal cancellation	Subtract involuntary signal	[Sirouspour and Salcudean, 2003]	No active CD required	Operator excluded (no shared control)

the human operator from its role as BDFT mitigator. By implementing a suitable vehicle-based mitigation method, BDFT can be
avoided, reduced or canceled regardless of the neuromuscular setting of the human operator. In the following, it is discussed which
of the vehicle-based approaches are considered most promising. Recall the four requirements proposed in Section 6.1, stating that a
promising BDFT solution should relieve the human operator from
its role as BDFT mitigator, should be generally applicable across
different vehicles, should require only limited modifications to the
vehicle, and should not purely rely on the separation assumption.

First, as we are considering vehicle-based solutions, we can remove
neuromuscular adaptation (#3) from Table 6.1 as preferred solution
for BDFT problems.

Then, as we are interested in a general solution to BDFT, applicable
to virtually any vehicle, we can discard the approaches that do not
fulfill this requirement, being: minimize PLF motion (#1), HO-CD
restraints (#4.2) and axes design (#5.3). Also the cabin vibration reduction method, proposed in [Humphreys et al., 2011], falls in this
category, as it is only applicable to excavator(-like) systems. Also,
let us discard the approach of active PLF-HO isolation (#2.3) due
on the many practical complications it involves, which limit the applicability of this approach.

Then, the approaches that rely on the separation assumption are
discarded, as this assumption is questionable. The approaches that
are to be discarded based on this observation are the CD responsiveness (#5.1) and the filtering approaches (#6.1 and #6.2). Finally,
adaptations to the control device dynamics (#5.2), such as increasing its stiffness, can be disregarded as general solution due to the
negative effects on controllability it often introduces.

Two general approaches remain as promising general BDFT mitigation techniques. The first approach consists of passive measures to
restrain and immobilize body parts (#2.1, #2.2 and #4.1), which are
already commonly applied. Note that this approach includes two
of the seven solution types, i.e., PLF-HO interface design and HO-
CD interface design, which we group together here for convenience.
The second promising approach is the model-based cancellation approach, where use is made of a BDFT model to obtain a canceling

signal (#7.1 and #7.2). Successful model-based cancellation relies heavily on the accuracy of this model and the development of such a model is a challenging task. However, once such a model is obtained, it is accompanied by an increased understanding of BDFT, which can be regarded as an additional advantage of pursuing the model-based solution approach.

In the following, the potential of the two selected mitigation approaches will be discussed in more detail. As there are several possible embodiments for both approaches, a choice was made for each of them. As an example of the passive support/restraining systems an *armrest* was selected (which belongs to HO-CD interface design approaches, #4.1). For the second approach, model-based *signal cancellation* was selected as topic for further investigation (#7.2). Evidently, this selection leaves some options, such as model-based force cancellation, unexplored. As not all available options can be covered within the context of this thesis, these are left for future investigation.

6.4   Potential of armrest in BDFT mitigation

Only few works have been devoted to the effect of an armrest on BDFT effects. This is surprising, especially considering the fact that an armrest is a straightforward, low-cost and potentially very effective tool in mitigating biodynamic feedthrough. In the extensive review of the effects of translational whole-body vibration on continuous manual control performance by McLeod and Griffin [1989], it is remarked that "the only study to investigate the effect of providing an arm-support on performance was conducted by Torle" (p. 66), referring to [Torle, 1965].

In [Torle, 1965] it was shown that an armrest resulted in greater improvement in tracking performance under vertical gust accelerations than could be obtained by optimizing other control device parameters. In an experiment subjects were exposed to vertical gust-like disturbances, while performing a control task. Two different armrests were used, a board of 25 cm in length and a cross-piece of 4 cm in width. In addition, adaptations of the control device dynamics were studied: various amounts of backlash and different levels of friction. Backlash was found to cause a deterioration in tracking performance and no advantage of including friction was observed. Both armrest significantly improved control performance (no difference was found between the two armrest configurations). In [McLeod and Griffin, 1995] it was remarked that the addition of an armrest reduced the magnitude of vibration breakthrough occurring at the control, particularly for 4 Hz vibration. As a possible explanation of the effect of the armrest, it was suggested that an armrest may reduce the relative movement between hand and elbow during exposure to low frequency vibration, reducing the neuromuscular interference (the phenomenon where vibrations affect the feedback mechanisms in a limb, causing perceptual confusion about the forces generated by and the positions of a limb) [McLeod and Griffin, 1995]. Despite these promising results, the effect of an armrest on BDFT dynamics received limited further attention.

The results of a study into the effect of an armrest on BDFT dynamics, performed within the context of this thesis, will be presented in Chapter 7. Already before measuring and presenting the results, some remarks regarding the expected results can be made.

An armrest would be expected to alter the dynamics between human operator and control device, hence influencing H_{HOCD} dynamics. In Fig. 6.1 the expected effect of an armrest is indicated at #4. It is to be expected that the addition of an armrest reduces BDFT effects by stabilizing the arm.

The armrest may also have an influence on the admittance estimates and force disturbance feedthrough (FDFT) estimates, as the limb's response to force disturbances may be altered. It is important to note that the limb itself is not changed by the addition of the armrest, but that effects like stabilization of the arm and changes in muscular activation (such as pressing the arm against the armrest) may impact the overall limb dynamics and, e.g., reduce the position deviation that results from a force disturbance.

As the admittance estimate and FDFT estimate are based on this relationship the admittance and FDFT estimates may change. Whether and to which extent an armrest influences admittance and FDFT

dynamics is likely to depend on both the configuration and location of the armrest. The response to force disturbances involves
mainly forearm dynamics and rotations around the elbow. Hence,
an armrest located close to the control device (supporting the forearm close to the wrist, limiting forearm motions) is likely to have
a stronger reducing effect on force disturbance responses than one
that is further away from the control device (supporting the forearm close to the elbow, leaving the forearm free to swing). For
the BDFT dynamics, the location of the armrest is likely to be beneficial in both cases as they depend on involuntary arm motions
in which both lower and upper arm are involved. An armrest far
away from the control device, supporting the elbow, reduces the
involuntary swinging of the upper arm and so reduces the involuntary inputs. An armrest close to the control device, supporting the
wrist, leaves the forearm largely free to swing, but reduces the involuntary excursions of the wrist, also reducing involuntary inputs.
It is possible that the effect of the armrest on BDFT dynamics depends on the setting of the neuromuscular system. Chapter 7 will
further investigate the speculations provided here.

6.5   Potential of model-based cancellation

In the following the potential of the second type of promising BDFT
mitigation approaches will be evaluated: model-based cancellation.
The focus will be on signal cancellation.

6.5.1   Method of evaluation: Optimal Signal Cancellation

In model-based cancellation a BDFT model is used to estimate the
involuntary part in the control inputs. In signal cancellation a
model is used that calculates involuntary control device deflections
using the PLF acceleration as input. Hence, the model describes
biodynamic feedthrough to positions (B2P) dynamics. We will call
this model H_{B2P}^{mod}. Its output is then used to remove the BDFT influence, simply by subtracting it from the total control input before it
enters the CE, illustrated as signal cancellation (SC) in Fig. 6.1.
Different types of models can be used for signal cancellation: e.g.,

it could be a physical model, a black box model, or even the frequency response directly (which will be used for this evaluation). In any case, the model should describe the transfer of accelerations to involuntary control device deflections. *Optimal* signal cancellation, as introduced in [Venrooij et al., 2011b], is an offline technique which allows for studying the potential of signal cancellation. Under certain circumstances, the method allows for complete cancellation of BDFT effects. This is achieved by using the BDFT frequency response estimate as a model. Complete cancellation is only guaranteed if the model is applied to the same data that was used to construct the model. Optimal signal cancellation is only possible 'offline' and not suitable as a practical solution. The method is used here as it meets our current goals in providing insight in the potential of signal cancellation as a BDFT mitigation technique.

6.5.2   Experiment: measuring biodynamic feedthrough

The experiment that was performed to obtain the data used for this study was already described in Chapter 4. For a complete description of the experiment, the reader is referred to that chapter.

In summary, twelve subjects participated in an experiment in the TU Delft's SIMONA Research Simulator. The control device was an electrically actuated side-stick, located at the right-hand side of the subject. No armrest was present. Only lateral (left-right) acceleration disturbances were studied. In the experiment, subjects performed three disturbance rejection tasks [Abbink, 2006]: (i) position task (PT), or 'stiff task', in which the instruction was to keep the position of the side-stick centered, that is, to "resist the force perturbations as much as possible"; (ii) force task (FT), or 'compliant task', in which the instruction was to minimize the force applied to the side-stick, that is, to "yield to the force perturbations as much as possible"; and (iii) relax task (RT), in which the instruction was to relax the arm while holding the side-stick, that is, to "passively yield to all perturbations". For the PT the best performance is achieved by being very stiff, the FT requires the operator to be very compliant. The RT yields an admittance reflecting the passive dynamics of the neuromuscular system. In earlier studies, identical tasks were used and it was shown that the BDFT dynamics strongly depend on these control tasks (see Chapter 2).

During the experiment, a motion disturbance $M_{dist}(t)$ was applied to the simulator's motion base – in order to measure the BDFT – and a force disturbance $F_{dist}(t)$ was applied to the side-stick – in order to measure the neuromuscular admittance. Due to the presence of the two disturbance signals, the measured control device deflection consisted of three distinct contributions:

$$\theta_{cd}(t) = \theta_{cd}^{Fdist}(t) + \theta_{cd}^{Mdist}(t) + \theta_{cd}^{res}(t) \qquad \text{6.1}$$

where the superscript ^{Fdist} denotes the contribution of the force disturbance and ^{Mdist} the contribution of the motion disturbance (the biodynamic feedthrough). The remaining part of the control input signal, i.e., the part that is *not* related to any of the two disturbance signals, is labeled here the 'residual' and denoted by the superscript ^{res}. The residual includes cognitive contribution from the CNS and the remnant.

The estimate of the B2P dynamics, is calculated using the estimated cross-spectral density between $M_{dist}(t)$ and $\theta_{cd}(t)$ ($\hat{S}_{mdist,\theta}(j\omega_m)$) and the estimated auto-spectral density of $M_{dist}(t)$ ($\hat{S}_{mdist,mdist}(j\omega_m)$):

$$\hat{H}_{B2P}(j\omega_m) = \frac{\hat{S}_{mdist,\theta}(j\omega_m)}{\hat{S}_{mdist,mdist}(j\omega_m)} \qquad \text{6.2}$$

where ω_m are the frequencies of the motion disturbance signal $M_{dist}(t)$ and j is the imaginary unit. Also the squared coherence, a measure for the dynamics' linearity and reliability of the estimate, was calculated:

$$\hat{\Gamma}^2_{B2P}(j\omega_m) = \frac{\left|\hat{S}_{mdist,\theta}(j\omega_m)\right|^2}{\hat{S}_{mdist,mdist}(j\omega_m)\hat{S}_{\theta,\theta}(j\omega_m)} \qquad \text{6.3}$$

The squared coherence equals one when there are no non-linearities and no time-varying behavior and zero when there is no linear behavior at all.

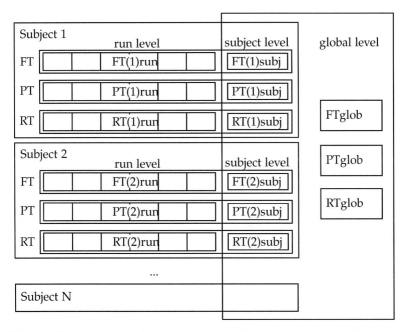

Figure 6.3: Experimental data can be viewed at run level, subject level and global level for each task.

6.5.3 Results: biodynamic feedthrough data

The results of the aforementioned experiment can be presented in different levels of 'generality', ranging from specific, i.e., the result of a single run of a single subject, to global, i.e., the average result over all runs of all subjects. The different levels of generality and their relation are shown in Fig. 6.3. The lowest level of generality is the result of a single run; this level is called the *run level*. By taking the average over several repetitions of a given experimental condition for a given subject, *subject level* data is obtained: a more general result, but still specific for one subject. The average of all repetitions of a certain condition over all subjects is called the *global level*.

Fig. 6.4 shows the B2P dynamics for different levels of generality:

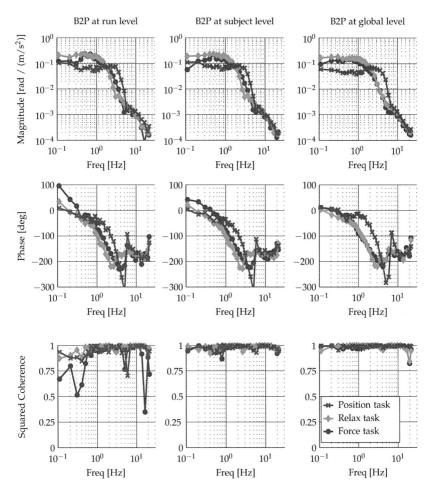

Figure 6.4: Biodynamic feedthrough results for different levels of generality for the three control tasks.

the left-hand column shows the results (magnitude, phase and co-herence) of the *run level* data of a selected (typical) subject. The mid-dle column shows the *subject level* data for the same subject. The right-hand column shows the average for each control task over all repetitions of all subjects (*global level*). From the results it can be observed that B2P is task dependent: different BDFT dynamics are measured for the different tasks. Furthermore, looking at the different generality levels, the plots show similar features but also several differences. By increasing the generality level, some fea-tures become more pronounced while other are removed. A clear difference between the generality levels is observed for the squared coherence: it increases for an increasing level of generality. This is in line with expectations as it is known that time-averaging over repetitions removes noise [Schouten et al., 2008a].

The results shown in Fig. 6.4 raise several questions. Looking at the different tasks, the B2P dynamics clearly depend on the task that is performed: how important are the differences? Looking at the different generality levels, the results look similar: does that mean we can generalize BDFT dynamics across subjects? How would a model based on global level data perform in signal can-cellation compared to a model based on subject or run level data? Which level of generality is optimal for use in the signal cancella-tion model? With optimal signal cancellation (OSC) these questions can be answered.

6.5.4 Methods: optimal signal cancellation

In the following, the term *data* refers to the measured response, that is, the measured control device angle $\theta_{cd}(t)$. The term *model* is used for the signal cancellation model that describes the transfer of ac-celerations to involuntary control device deflections. In this study the B2P estimate $\hat{H}_{B2P}$, obtained through Eq. 6.2, is used as model. This guarantees perfect cancellation of all BDFT effects, as long as the model is applied to the same data that was used to construct the model.

Optimal signal cancellation is performed as follows: by multiplying the estimated B2P dynamics ($\hat{H}_{B2P}$) with the auto-spectral density of the motion disturbance signal ($S_{mdist,mdist}$) the cross-spectral density of $M_{dist}(t)$ and $\theta_{cd}(t)$ is obtained:

$$\hat{S}_{mdist,\theta}(j\omega_m) = \hat{H}_{B2P}(j\omega_m)\hat{S}_{mdist,mdist}(j\omega_m) \qquad \text{6.4}$$

From this, the Fourier transform of the control device deflection estimate can be calculated using:

$$\theta_{cd}^{mod}(j\omega_m) = \frac{\hat{S}_{mdist,\theta}(j\omega_m)N^2}{M_{dist}^*(j\omega_m)} \qquad \text{6.5}$$

were N is the number of samples in the time signal and $M_{dist}^*(j\omega_m)$ is the complex conjugate of the Fourier transform of the motion disturbance signal. By inverse-Fourier transforming $\theta_{cd}^{mod}(j\omega_m)$, a time signal $\theta_{cd}^{mod}(t)$ is obtained. This signal is the *model output*: an estimate of involuntary control device deflections, caused by biodynamic feedthrough, based on the BDFT dynamics in the model and the acceleration signal in the data. By subtracting this signal from the measured control device deflection, the signal cancellation result is obtained, $\theta_{can}(t)$. This is the 'corrected' control device deflection angle after the cancellation. This signal contains contributions from the two disturbances, $F_{dist}(t)$ and $M_{dist}(t)$, the residual and the effect of signal cancellation:

$$\theta_{can}(t) = \theta_{cd}^{Fdist}(t) + \theta_{cd}^{Mdist}(t) + \theta_{cd}^{res}(t) - \theta_{cd}^{mod}(t) \qquad \text{6.6}$$

Compare this with the original stick deflection, Eq. 6.1. If the model output $\theta_{cd}^{mod}(t)$ approximates $\theta_{cd}^{Mdist}(t)$, the effects of motion disturbances on the control device deflections are removed.

The success of the cancellation is evaluated using frequency decomposition (see Chapter 2). This method decomposes the total uncorrected stick deflection $\theta_{cd}(t)$ in three parts: the contributions of $F_{dist}(t)$, the contribution of $M_{dist}(t)$ and a residual. By comparing the actual contribution of the motion disturbance $\theta_{cd}^{Mdist}(t)$ with the modeled one $\theta_{cd}^{mod}(t)$, the accuracy of the cancellation can be evaluated. This success is quantified by calculating the root-mean-square (RMS) of the difference between $\theta_{cd}^{Mdist}(t)$ and $\theta_{cd}^{mod}(t)$ and dividing this difference by the RMS of $\theta_{cd}^{Mdist}(t)$. The result is expressed as

Table 6.2: BDFT cancellation (P_{can}) for (a) subject level models and (b) global level models. The table indicates the mean across subjects with the standard deviation in parenthesis.

<table>
<tr><td colspan="3">(a) Subject level models</td><td colspan="3">(b) Global level models</td></tr>
<tr><td></td><td>Mean</td><td>SD</td><td></td><td>Mean</td><td>SD</td></tr>
<tr><td>PT</td><td>66.40%</td><td>(6.4%)</td><td>PT</td><td>51.95%</td><td>(8.7%)</td></tr>
<tr><td>FT</td><td>61.60%</td><td>(9.1%)</td><td>FT</td><td>43.65%</td><td>(10.1%)</td></tr>
<tr><td>RT</td><td>81.64%</td><td>(6.1%)</td><td>RT</td><td>47.62%</td><td>(26.9%)</td></tr>
</table>

P_{can}: a percentage which indicates the reduction in RMS of the M_{dist} part after cancellation as a percentage of the original RMS value:

$$P_{can} = \left(1 - \frac{RMS\left(\theta_{cd}^{Mdist}(t) - \theta_{cd}^{mod}(t)\right)}{RMS\left(\theta_{cd}^{Mdist}(t)\right)}\right) \qquad \boxed{6.7}$$

A result of 100% BDFT cancellation means perfect cancellation, a result between 0% and 100% means a partial reduction of BDFT effects, 0% means no reduction at all, and a negative result means an increase in BDFT effects. By combining different models and data the accuracy of a model can be tested.

In this chapter the OSC results in three cases are studied:

- *Subject level models*: how well can a subject level model describe run level data?

- *Global level models*: how well can a global level model describe run level data?

- *Across tasks*: how well does a BDFT model of one task describe the BDFT data of other tasks?

6.5.5   Results

Subject level models

The result of using OSC with subject level models on run level data is shown in Table 6.2-A and Fig. 6.5. The results were calculated by averaging the percentage of BDFT cancellation for each task over

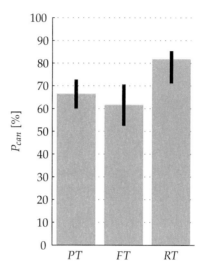

Figure 6.5: BDFT cancellation for subject level models. The bars indicates the mean across subjects and the lines the standard deviation.

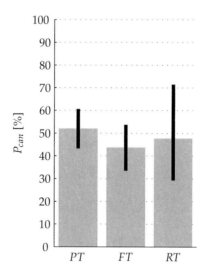

Figure 6.6: BDFT cancellation for global level models. The bars indicates the mean across subjects and the lines the standard deviation.

all runs of all subjects.

The data show that with a subject level model, applied to run data, between 60% to 80% of the BDFT effects can be canceled. The best results are obtained for the relax task. This implies that a subject level model describes the behavior of individual runs well. Whether the model quality is sufficient depends on the intended purpose. In general, however, it can be said that using a subject level model on run level data provides substantial cancellation.

Global level models

The result of using OSC with global level models on run level data is shown in Table 6.2-B and Fig. 6.6. The results show the level of BDFT cancellation for each task, averaged over all runs of all subjects. The data show that with a global level model, applied to run data, between 40% to 50% of the BDFT effects can be canceled. This is lower than the result obtained for the subject level models, which implies that by averaging over the subjects the applicability of the BDFT model for cancellation is reduced, as could be logically expected.

Nonetheless, the amount of cancellation that is achievable with global level models is still considerable. Although better results can be obtained with a more specific model, a global model can still be used to reduce biodynamic feedthrough effects by roughly 50%. It should be kept in mind, however, that the subjects in this study belonged to a homogeneous group of right-handed male subjects with only a small variation in age, mass and height. It is to be expected that results differ for different groups of subjects. With the current data set, however, the influence of body type on the results cannot be investigated.

In conclusion, signal cancellation with global level models is possible, but is outperformed by using subject level models. A similar observation was made for 'global scope' and 'individual scope' models in Chapter 5.

Table 6.3: BDFT cancellation (P_{can}) across tasks. The table indicates the mean across subjects and the standard deviation between parenthesis.

		Model		
		PT	FT	RT
	PT	100.0% (0.0%)	-75.2% (44.7%)	-116.1% (81.7%)
Data	FT	9.6% (10.1%)	100.0% (0.0%)	24.4% (30.3%)
	RT	12.2% (9.4%)	38.5% (23.3%)	100.0% (0.0%)

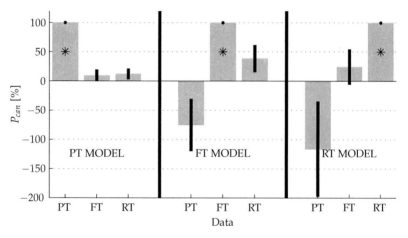

Figure 6.7: BDFT cancellation across tasks. The asterisks indicate the cases where model and data were matched. The bars indicates the means across subjects and the lines the standard deviations.

Across tasks

The result of using OSC with subject level models across tasks is shown in Table 6.3 and Fig. 6.7. The results show the level of BDFT cancellation for the subject level model of each task applied to the subject level data of each task, averaged over all runs of all subjects. The results show, first of all, that when the model of a task is applied to that data of that task, 100% cancellation is achieved (marked with asterisks in the figure). This is because optimal signal cancellation is designed such that perfect cancellation is achieved when model and data are matched. Secondly, when attempting cancellation across tasks, a poor cancellation result is obtained. The best results are obtained when applying a force task model to relax task data, or vice-versa. This is in line with expectations, as we already observed in Fig. 6.4 that these two tasks have quite similar BDFT dynamics. The worst results are obtained for PT data. Both the FT model and RT model result in a negative value for P_{can}, indicating that the cancellation deteriorates the situation rather than improving it. That implies that the modeled signal matches so poorly with the actual dynamics that the cancellation fails completely. The 'safest' model choice appears to be the position task model, as it shows a partial reduction, even across tasks. But the level of cancellation is only marginal.

As was shown in Chapter 2, the changes in BDFT dynamics across tasks are caused by the adaptations of neuromuscular dynamics of the human operator. In the experiment the subjects changed their neuromuscular dynamics and by doing so they also changed the BDFT dynamics. So when we say that biodynamic feedthrough models should be task dependent, it implies that the models should be capable of adapting in agreement with changes in the neuromuscular dynamics of the human operator. From the aforementioned results it becomes clear that adequate cancellation only occurs when the model matches the task, i.e., when the model matches the current setting of the neuromuscular dynamics.

6.6   Conclusions

Using a general BDFT system model, the available BDFT mitigation techniques were listed and evaluated. In total, seven different solution types, each providing one or more solution approaches, were identified and discussed. After discarding the solution types that do not meet requirements on general applicability and allowable complexity or rely on the questionable separation assumption, two solution types remain that are deemed most promising. Measures of the first solution type – passive support/restraining systems (e.g., seat belts and armrests) – are already commonly applied. Studies have shown that these are not sufficient to remove BDFT completely. The second solution type is model-based BDFT cancellation, where use is made of a BDFT model to obtain a prediction that is used as canceling signal. This approach has received some attention, but only very few experimental implementations have been tested.

A promising implementation of a passive support/restraining systems is the armrest. Surprisingly, only very few studies have investigated the effect of armrests on BDFT dynamics. Using the BDFT framework, developed in Chapter 3, several hypotheses were proposed regarding the influence of an armrest on both force and motion disturbance responses.

The potential of signal cancellation was investigated using a method called optimal signal cancellation. The signal cancellation results were evaluated for different levels of generality and for different tasks.

The results show that using subject level models, applied to run level data, between 60% to 80% of the BDFT effects can be canceled. When using global level models the cancellation reduces to a values between 40% and 50%. This implies that averaging over the subjects reduces the applicability of the BDFT model for BDFT cancellation. Signal cancellation with global level models is possible, but they are outperformed by subject level models. When attempting signal cancellation across tasks, it is observed that good cancellation can only be achieved when the model matches the task.

The current study indicates that signal cancellation is only a promising mitigation method for BDFT problems if the model can be adapted to both subject and task. Adaptation to task, or more correctly, to the neuromuscular dynamics of the human operator is of particular importance. A failure to identify changes in the neuromuscular settings of the human operator and adapting the model accordingly leads to suboptimal or incorrect control actions. The results show that reliable and fast online identification method of the neuromuscular dynamics of the human operator are a requirement for real-time adaptive BDFT signal cancellation.

CHAPTER 7

Mitigating biodynamic feedthrough with an armrest

In this chapter the BDFT mitigation effectiveness of a simple, cheap and widely-used hardware component is studied: the armrest. An experiment was conducted in which the BDFT dynamics were measured with and without armrest for different levels of neuromuscular admittance. The results show that the effect of the armrest on BDFT dynamics varies, both with frequency and neuromuscular admittance. It can be concluded that an armrest is an effective tool in mitigating BDFT.

The contents of this chapter are based on:

Paper title   How Effective is an Armrest in Mitigating Biody-
                  namic Feedthrough?

Authors   Joost Venrooij, Mark Mulder, Marinus M. van
                  Paassen, David A. Abbink, Frans C. T. van der
                  Helm, Max Mulder, and Heinrich H. Bülthoff

Published at 2012 IEEE International Conference on Systems,
                  Man and Cybernetics, October 14-17, 2012, Seoul,
                  Korea

7.1   Introduction

In this chapter the biodynamic feedthrough (BDFT) mitigation effectiveness of a simple, cheap and widely-used hardware component is studied: the armrest. Surprisingly little is known about the effect of arm support on BDFT dynamics. Armrests can be found in a wide range of vehicles – from fighter jets to electric wheelchairs – and are often primarily employed to increase steering comfort and reduce fatigue. However, they are also effective in stabilizing the arm when subjected to motion disturbances. Some studies suggested that an armrest may decrease the level of biodynamic feedthrough [McLeod and Griffin, 1989; Sövényi, 2005], but often did not answer the questions '*how?*' and '*by how much?*'. An exception is a study by Torle [1965], where the effect of an armrest was investigated by evaluating the tracking performance in an environment with random vertical accelerations with and without armrest. It was found that by providing an armrest, a greater improvement in tracking performance was achieved than by optimizing other control-stick parameters. Another study that investigated the effect of an armrest on BDFT effects is [McLeod and Griffin, 1995]. In that study, where vibration disturbances of 0.5 Hz and 4 Hz were used, it was observed that the addition of an armrest reduced the magnitude of BDFT effects, particularly for 4 Hz vibration. As a possible explanation of the effect of the armrest, it was suggested that an armrest may reduce the relative movement between hand and elbow during exposure to low frequency vibration, reducing the neuromuscular interference (the phenomenon where vibrations affect the feedback mechanisms in a limb, causing perceptual confusion about the forces generated by and the positions of a limb) [McLeod and Griffin, 1995]. Even though the results showed that the presence of an armrest did not remove *all* BDFT effects, the armrest – being a cheap component that is easily designed and installed – has the potential of being an excellent BDFT mitigation tool.

The BDFT system model is shown in Fig. 7.1. An extensive description of the BDFT system model was already provided in Chapter 3, which will not be repeated here. The effect of an armrest is indicated. An armrest will influence the H_{HOCD} dynamics by altering the feedthrough of accelerations through the human body into involuntary forces.

It is known that humans can and do vary the neuromuscular admittance of their limbs, e.g. their limb dynamics (H_{NMS} in Fig. 7.1). A human can, for example, depending on the task at hand, change his limb dynamics from being 'compliant' to being 'stiff'. The results in Chapter 2 showed that BDFT dynamics change when the limb dynamics of the operator change. When investigating the effect of an armrest on the BDFT dynamics, this variability of the BDFT dynamics due to neuromuscular admittance needs to be taken into account. The approach of the current study is thus to determine the effectiveness of an armrest as a BDFT mitigation tool by measuring the BDFT dynamics for several subjects with and without armrest for three levels of the neuromuscular admittance, varying between 'compliant', 'relaxed' and 'stiff' dynamics.

The experiment that was conducted is presented in Section 7.2. The analysis is discussed in Section 7.3, followed by the results in Section 7.4. The chapter ends with conclusions in Section 7.5.

7.2   Experiment

An experiment was designed in which the BDFT dynamics were measured with and without armrest for different levels of neuromuscular admittance. The experiment was performed in open-loop. This implies that the human operator did not have an influence on the acceleration signal, as opposed to the closed-loop BDFT system where the operator is influencing the accelerations through control inputs. For experimental purposes, an open-loop BDFT system is preferred over the closed-loop type because it allows the experimenter to design the acceleration signal. One can assume that the armrest effects studied here in open-loop are similar to those occurring in the closed-loop cases, i.e., in actual vehicles.

The experiment presented in Chapter 4 provided the data without armrest. An additional experiment was performed with an armrest. Except for the presence of the armrest, this experiment was

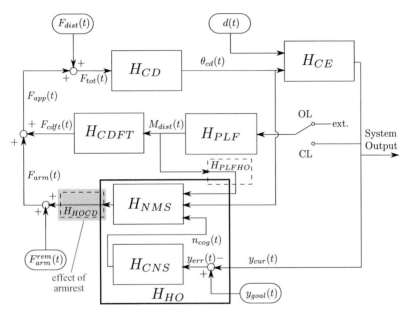

Figure 7.1: The biodynamic feedthrough system model. A human operator (HO) controls a controlled element (CE) using a control device (CD). Motion disturbances $M_{dist}(t)$ are coming from the platform (PLF). The feedthrough of $M_{dist}(t)$ to involuntary applied forces $F_{arm}(t)$ and involuntary control device deflections $\theta_{cd}(t)$ is called biodynamic feedthrough (BDFT). The feedthrough of $M_{dist}(t)$ to inertia forces $F_{cdft}(t)$ is called control device feedthrough (CDFT). $F_{app}(t)$ is the sum of the forces applied to the control device by the HO. The HO consists of a central nervous system (CNS) and a neuromuscular system (NMS). The connection between the HO and the environment is governed by two 'interfaces', H_{PLFHO} and H_{HOCD}. The CE and PLF can form an open-loop (OL) or closed-loop (CL) system. The focus of the current chapter: the influence of an armrest on force disturbance responses and motion disturbance responses is investigated. The armrest influences the H_{HOCD} dynamics.

Figure 7.2: The side-stick and armrest used in the experiment.

performed in an identical fashion. In the following, the two experiments will be described briefly; for more details the reader is referred to Chapter 4.

In this study, the influence of an armrest is determined using two types of dynamics: the force disturbance responses – i.e., the admittance and force disturbance feedthrough (FDFT) dynamics – and the motion disturbance responses – i.e., the BDFT to positions (B2P) and BDFT to forces (B2F) dynamics.

7.2.1   Apparatus

The experiment was performed in the TU Delft's SIMONA Research Simulator (SRS). The control device was an electrically actuated side-stick, located at the right-hand side of the subject. The measurements were performed with a side-stick because an armrest seems to be most practical for this type of control device. The armrest was a fixed (but removable) platform on which the subject could rest his arm (see Fig. 7.2). During the experiment, a motion disturbance $M_{dist}(t)$ was applied to the simulator's motion base – in order to measure the BDFT – and a force disturbance $F_{dist}(t)$ was applied to the side-stick – in order to measure the neuromuscular admittance. Only lateral (left-right) acceleration disturbances were

Table 7.1: Data of subjects (N = 7, all male and right-handed).

	Age [years]	Weight [kg]	Height [cm]	Fore arm [cm]	Upper arm [cm]	BMI [kg/m²]
mean	25.0	73.7	185.0	28.9	35.3	21.6
st. dev.	1.6	4.5	7.7	1.1	2.4	2.3
range	22-27	68-82	175-194	28-30	28-30	18.1-24.5

studied. Also the neuromuscular admittance was only measured in lateral direction by using a lateral force disturbance on the control device. The control device was fixed in the longitudinal direction. A head-down display was located in front of the subject. The seat had a five-point safety belt that was adjusted tightly in the experiments, to reduce torso motions.

7.2.2 Subjects

Fourteen subjects participated in the experiment without armrest and fourteen subjects participated in the experiment with armrest. Seven subjects participated in both experiments. For the current analysis, only the data of these latter seven subjects will be analyzed. They were all male, right-handed and had a small variation in age (between 22 and 27 years) and body mass index (BMI) (between 18.1 and 24.5 kg/m²). Details on the subjects can be found in Table 7.1, listing the mean, standard deviation (SD) and range of subject parameters.

7.2.3 Task and task instruction

In the experiment, subjects performed three disturbance rejection tasks [Abbink, 2006]: (i) position task (PT), or 'stiff task', in which the instruction was to keep the position of the side-stick centered, that is, to "resist the force perturbations as much as possible"; (ii) force task (FT), or 'compliant task', in which the instruction was to minimize the force applied to the side-stick, that is, to "yield to the force perturbations as much as possible"; and (iii) relax task (RT), in which the instruction was to relax the arm while holding the side-stick, that is, to "passively yield to all perturbations". For the PT the best performance is achieved by being very stiff, the FT requires the operator to be very compliant. The RT yields an admittance reflect-ing the passive dynamics of the neuromuscular system. In earlier studies, identical tasks were used and it was shown that the BDFT dynamics strongly depend on these control tasks (see Chapter 2).

Before entering the simulator, subjects were instructed on the goal of the experiment and the control tasks they were to perform. Sev-eral training runs were conducted to allow subjects to get used to the disturbances and the control tasks. During training, visual per-formance feedback was provided on the screen to help the subject understand the differences between the control tasks (see Chap-ter 4). After the execution of the task a score was provided. When a consistent performance (i.e., score) was reached, the visual per-formance feedback was removed from the screen for the remainder of the experiment – to minimize cognitive control actions based on visual feedback – and the actual measurement started.

7.2.4   Independent variables

Two independent variables were used: the control task (TSK) and the presence of an arm support in the form of an armrest (AR). The different levels were:

1. TSK: Position task (PT); Force task (FT); and Relax task (RT)

2. AR: arm support present in the form of an armrest (SUP); and no arm support present (NOSUP)

Together, TSK and AR yielded six different conditions. Each of the conditions was repeated six times. For all subjects, the NOSUP condition was measured first, followed by the SUP condition. The order of the control tasks was randomized.

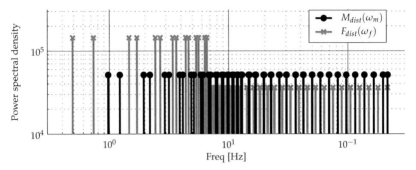

Figure 7.3: Power spectral density plot of disturbance signals F_{dist} and M_{dist}.

7.2.5 Perturbation signal design

Both disturbance signals $F_{dist}(t)$ and $M_{dist}(t)$ were multi-sines, defined in the frequency domain. By separating the signals in frequency, see Fig. 7.3, the response due to each disturbance could be identified in the measured signals (Chapter 2). To obtain a full bandwidth estimate of the admittance, a range of frequencies, ω_f, between 0.05 Hz and 21.5 Hz was selected for the force disturbance signal $F_{dist}(t)$. For the motion disturbance signal $M_{dist}(t)$, a range of frequencies, ω_m, between 0.1 and 21.5 Hz was selected.

7.2.6 Perturbation signal scaling

To minimize the effect of non-linearities and to be able to compare between different control tasks, the stick deflection in each task should be small and similar in size. In an effort to achieve this, the disturbance signals were scaled in a tuning procedure [Venrooij et al., 2011a]. The goal of this procedure was to find, for each control task, the gains needed to result in similar standard deviation (3 degrees) of the control device deflection $\theta_{cd}(t)$. The tuning was done in the condition without armrest. The scaling factors for $F_{dist}(t)$ found for the PT, FT and RT were 20, 1 and 0.38 respectively. For the $M_{dist}(t)$ signal, the scaling factors for PT, FT and RT were 0.9, 0.7 and 0.7 respectively.

7.3 Analysis

7.3.1 Calculating the dynamics

The control device deflection, $\theta_{cd}(t)$, and the applied force to the control device, $F_{app}(t)$, were averaged in the time domain over the six repetitions of a condition.

Force disturbance responses

The admittance was estimated in the frequency domain, with a closed loop identification technique that used the estimated cross-spectral densities between $F_{dist}(t)$ and $\theta_{cd}(t)$ ($\hat{S}_{fdist,\theta}(j\omega_f)$) and between $F_{dist}(t)$ and $F_{arm}(t)$ ($\hat{S}_{fdist,f}(j\omega_f)$) (see also Chapter 3):

$$\hat{H}_{adm}(j\omega_f) = \frac{\hat{S}_{fdist,\theta}(j\omega_f)}{\hat{S}_{fdist,f}(j\omega_f)} \tag{7.1}$$

The FDFT dynamics are calculated using the force disturbance signal $F_{dist}(t)$ and the control device deflections $\theta_{cd}(t)$. So, the FDFT estimate can be calculated, at frequencies ω_f, as follows:

$$\hat{H}_{FDFT}(j\omega_f) = \frac{\hat{S}_{fdist,\theta}(j\omega_f)}{\hat{S}_{fdist,fdist}(j\omega_f)} \tag{7.2}$$

Motion disturbance responses

The biodynamic feedthrough to positions (B2P) dynamics reflect the amount of involuntary control device deflections that occur under influence of an acceleration disturbance. The estimate of the B2P dynamics is calculated using the estimated cross-spectral density between $M_{dist}(t)$ and $\theta_{cd}(t)$ ($\hat{S}_{mdist,\theta}(j\omega_m)$) and the estimated auto-spectral density of $M_{dist}(t)$ ($\hat{S}_{mdist,mdist}(j\omega_m)$):

$$\hat{H}_{B2P}(j\omega_m) = \frac{\hat{S}_{mdist,\theta}(j\omega_m)}{\hat{S}_{mdist,mdist}(j\omega_m)} \tag{7.3}$$

In addition to the involuntary control device deflections, acceleration disturbances also cause involuntary forces applied to the control device. These can be represented using two different dynamics,

BDFT to forces in closed-loop (B2FCL) and BDFT to forces in open-loop (B2FOL). The difference between the two types of dynamics is which force signal is used as output signal (see Chapter 3 for details).

The B2FCL dynamics are calculated between $M_{dist}(t)$ as input and $F_{arm}(t)$ – i.e., the force applied to the control device by the operator's arm – as output. These dynamics can be estimated as follows:

$$\hat{H}_{B2FCL}(j\omega_m) = \frac{\hat{S}_{mdist,f}(j\omega_m)}{\hat{S}_{mdist,mdist}(j\omega_m)} \qquad \boxed{7.4}$$

It should be noted that the force $F_{arm}(t)$ is influenced by several signals. One of these is motion disturbance $M_{dist}(t)$, another is control device deflection $\theta_{cd}(t)$ (see Fig. 7.1). As was already noted in Chapter 3, this makes it hard to interpret B2FCL dynamics. We will see a demonstration of this when we look at the results of the B2FCL dynamics.

The B2FOL dynamics cannot be directly measured but can be obtained using the following relationship (see Chapter 3):

$$\hat{H}_{B2FOL}(j\omega_m) = \frac{\hat{H}_{B2P}(j\omega_m) - \hat{H}_{FDFT}(j\omega_m)H_{CDFT}(j\omega_m)}{\hat{H}_{FDFT}(j\omega_m)} \qquad \boxed{7.5}$$

where the $H_{CDFT}(j\omega_m)$ dynamics are the control device feedthrough. The CDFT dynamics can be defined as the transfer of accelerations through the control device mass, resulting in inertial forces being applied to the control device. These dynamics can be determined in a dedicated experiment. In our case the H_{CDFT} dynamics contain the mass of the control device (0.39 kg).

After having calculated the results for each condition and for each subject, the results were again averaged, now in the frequency domain and across subjects, to obtain one estimate for each type of dynamics per condition. These are the results that will be analyzed in the following.

7.3.2 Effect of the armrest

The B2P estimate obtained with and without the armrest is denoted as $\hat{H}_{B2P}^{sup}(j\omega_m)$ and $\hat{H}_{B2P}^{nosup}(j\omega_m)$, respectively. By comparing the dynamics obtained with and without the armrest, the influence of the armrest can be determined. An insightful way of visualizing differences is by looking at the ratio between the two dynamics. Ratio function $RF_{B2P}(j\omega_m)$ is defined as:

$$RF_{B2P}(j\omega_m) = \frac{\hat{H}_{B2P}^{sup}(j\omega_m)}{\hat{H}_{B2P}^{nosup}(j\omega_m)} \qquad \text{7.6}$$

such that:

$$\hat{H}_{B2P}^{sup}(j\omega_m) = RF_{B2P}(j\omega_m)\hat{H}_{B2P}^{nosup}(j\omega_m) \qquad \text{7.7}$$

In this way, the ratio function describes the effect of the armrest on each frequency point. A value $RF = (1 + 0j)$, i.e., a magnitude $|RF| = 1$ and a phase $\angle RF = 0$, means that the dynamics with and without armrest are equal on that frequency. Other values indicate a difference in magnitude or phase or both. Specifically, $|RF| > 1$ means the addition of an armrest caused an increase in magnitude, i.e., an increase in involuntary deflections, $|RF| < 1$ implies a decrease in magnitude, i.e., a decrease in involuntary deflections. A nonzero phase indicates that the armrest causes a shift in time, e.g., a delay, in the response to acceleration disturbances. In this study only the magnitude effects are investigated, as these are of greater importance when evaluating the potential of an armrest as tool to mitigate BDFT.

Similarly, a ratio function can be calculated for the admittance estimate:

$$RF_{adm}(j\omega_f) = \frac{\hat{H}_{adm}^{sup}(j\omega_f)}{\hat{H}_{adm}^{nosup}(j\omega_f)} \qquad \text{7.8}$$

the FDFT estimate:

$$RF_{FDFT}(j\omega_f) = \frac{\hat{H}_{FDFT}^{sup}(j\omega_f)}{\hat{H}_{FDFT}^{nosup}(j\omega_f)} \qquad \text{7.9}$$

and the two B2F estimates:

$$RF_{B2FCL}(j\omega_m) = \frac{\hat{H}^{sup}_{B2FCL}(j\omega_m)}{\hat{H}^{nosup}_{B2FCL}(j\omega_m)} \qquad \text{7.10}$$

$$RF_{B2FOL}(j\omega_m) = \frac{\hat{H}^{sup}_{B2FOL}(j\omega_m)}{\hat{H}^{nosup}_{B2FOL}(j\omega_m)} \qquad \text{7.11}$$

Each ratio function indicates the effect of the armrest on the dynamics.

7.3.3 Expected results

Now we have defined the ratio functions, it may be insightful to briefly discuss what we would expect to observe for the different ratio functions. Recall that in Chapter 6 some hypotheses were postulated regarding the expected influence of an armrest on force disturbance responses and motion disturbance responses. They are summarized here:

- It is to be expected that the addition of an armrest reduces BDFT effects by stabilizing the arm.

- The armrest may also have an influence on the force disturbance responses (admittance and FDFT). This influence likely depends on both the configuration and location of the armrest. The response to force disturbances involves mainly forearm dynamics and rotations around the elbow. Hence, an armrest located close to the control device (supporting the forearm close to the wrist) is likely to have a stronger reducing effect on force disturbance responses than one that is further away from the control device (supporting the forearm close to the elbow).

- For the motion disturbance responses (B2P, B2FOL, B2FCL), the addition of an armrest is likely to be beneficial independently of the location of the armrest, as these dynamics depend on involuntary arm motions in which both lower and upper arm are involved. An armrest supporting the elbow reduces the involuntary swinging of the upper arm; an arm-rest supporting the wrist reduces the involuntary excursions of the wrist.

• The effect of the armrest on motion disturbance responses may show a dependency on neuromuscular admittance.

7.4 Results

In the following the results of the experiment are presented. First, the effects on the force disturbance responses is addressed, followed by the effects on the motion disturbance responses.

7.4.1 Force disturbance responses

Fig. 7.4 shows the admittance, measured without and with armrest, and the ratio function. Only magnitude information is shown here, as it is most relevant for our current purposes. The lines indicate the means obtained by averaging over all subjects. The colored bands indicate the standard deviations (SD). The admittance measured without armrest shows the typical features: the lowest admittance is found for the PT; the highest for the FT; the admittance of the RT falls in between. The admittance measured with armrest shows very similar features. In fact, the most important thing to note is that the admittance data measured with and without armrest are actually very similar.

The ratio function allows for investigating the effect of the armrest in detail. The similarity in admittance measured with and without armrest is reflected by the ratio function, which shows several peaks but no structural increase or decrease in admittance. Only for the PT below 1 Hz a small but structural decrease in admittance can be observed ($|RF_{adm}| \approx 0.7$), meaning that the presence of the armrest allowed the subjects to express slightly stiffer behavior at lower frequencies. In general, it can be said that the admittance measured with and without armrest shows mainly unstructured differences. In other words, the presence of an armrest did not significantly impact the admittance of the human operator.

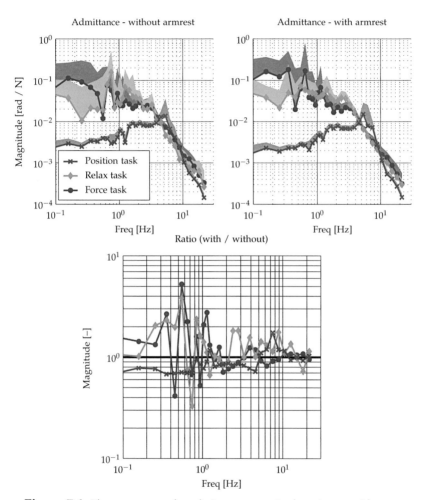

Figure 7.4: The neuromuscular admittance magnitude estimate, without armrest (left), with armrest (right) and the ratio function (bottom). The lines show means, the colored bands show standard deviation (mean + 1 SD).

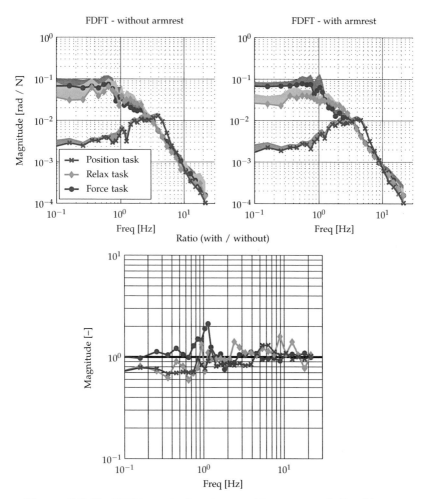

Figure 7.5: The FDFT magnitude estimate, without armrest (left), with armrest (right) and the ratio function (bottom). The lines show means, the colored bands show standard deviation (mean + 1 SD).

Fig. 7.5 shows the results for the FDFT measured without and with an armrest. The results corroborate and further clarify the observations made for the admittance. In Chapter 3 it was shown that the FDFT dynamics are the combination of the admittance and control device dynamics. The latter acts as a filter on the former, 'flattening' the dynamics in some cases. As a result the FDFT measurements are smoother than the admittance measurements, which provides a clearer picture, especially for the ratio function. The results confirm that the addition of the armrest did not have a large effect on the neuromuscular dynamics. The results suggest a small decrease in FDFT magnitude for the PT at low frequencies, as was also observed for the admittance. The results show a similar decrease in RT magnitude for low frequencies. The latter was not visible in the admittance data, most likely due to the relatively high amount of noise in the admittance data for this task. The results allow us to conclude that the addition of the armrest did not influence the admittance and FDFT dynamics to a large extent.

It was hypothesized that the influence of an armrest on the force disturbance responses may depend on the location of the armrest. The armrest used in this study was supporting the forearm mainly close to the elbow. It was already proposed that an armrest in such a location may only have limited influence on the admittance and FDFT dynamics, as these mainly depend on forearm dynamics and rotations around the elbow. The small reduction that was observed in the PT and RT magnitudes are in line with expectations, as for these tasks the armrest can be used to provide some extra stability for the arm. For the FT the instruction is to minimize the forces applied to the control device and in this case extra stability is not helpful. The results show that the operator's task performance was not affected by the addition of the armrest as the ratio function magnitude is close to 1 across the whole frequency range, indicating that the dynamics measured with and without armrest were similar.

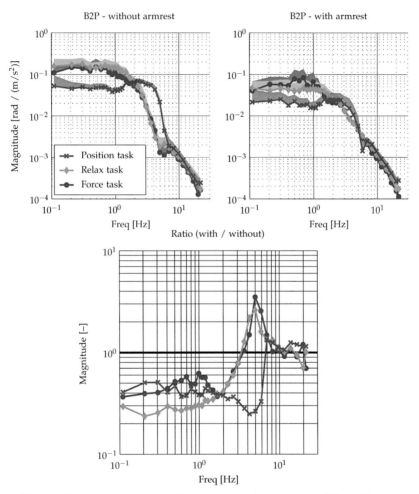

Figure 7.6: The B2P magnitude estimate, without armrest (left), with armrest (right) and the ratio function (bottom). The lines show means, the colored bands show standard deviation (mean + 1 SD).

Table 7.2: B2P ratio function magnitude $|RF_{B2P}|$.

freq. [Hz]	FT	PT	RT	freq. [Hz]	FT	PT	RT
0.11	0.37	0.41	0.30	2.26	0.49	0.35	0.49
0.21	0.39	0.50	0.24	2.65	0.65	0.37	0.60
0.31	0.40	0.51	0.26	3.04	0.79	0.33	0.79
0.40	0.44	0.41	0.30	3.63	1.04	0.28	1.28
0.50	0.52	0.47	0.28	4.21	1.49	0.25	2.18
0.60	0.53	0.37	0.27	4.99	3.50	0.26	2.61
0.70	0.57	0.38	0.28	5.87	2.55	0.33	1.61
0.79	0.49	0.44	0.28	6.95	1.48	1.32	1.33
0.89	0.49	0.41	0.30	8.22	1.02	1.24	1.32
0.99	0.62	0.38	0.30	9.68	1.05	1.12	1.11
1.09	0.57	0.38	0.30	11.54	0.92	1.06	1.01
1.18	0.56	0.44	0.34	13.78	1.07	1.25	1.10
1.28	0.48	0.42	0.33	16.52	0.91	1.20	0.95
1.48	0.43	0.40	0.35	19.74	1.19	1.14	0.71
1.67	0.37	0.38	0.39	21.50	0.70	1.15	1.01
1.97	0.40	0.38	0.41				

7.4.2 Motion disturbance responses

Fig. 7.6 shows the magnitude plot of the B2P dynamics measured without armrest, with armrest and the ratio function. The B2P dynamics measured without armrest show a dependency on the control task. At lower frequencies, the lowest B2P occurs for the PT and the highest for the RT. Around approximately 2-3 Hz, the B2P for the PT peaks and is higher than for other tasks. The results obtained with the armrest also show a dependency of task, although the relationships have somewhat changed. It can be observed that, in general, the magnitude of the B2P measured with an armrest is lower than without an armrest; this holds for all three tasks.

Table 7.2 lists the ratio function's magnitude values. From both Table 7.2 and Fig. 7.6 it can be concluded that the influence of the armrest on the B2P dynamics depends on both control task *and* disturbance frequency. The results presented in [Schoenberger and Wilburn, 1973] already suggested that the effect of an armrest were likely to be frequency dependent, the results of the current study allow to investigate this dependency in detail. The dependency of

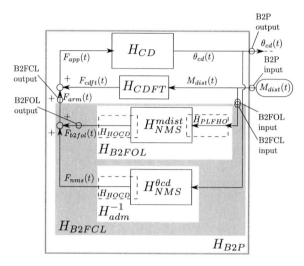

Figure 7.7: The B2P and B2FCL dynamics. The B2FCL dynamics can be split into two SISO systems, H_{B2FOL} and H_{adm}^{-1}, describing the B2FOL dynamics and inverse of the admittance respectively.

the effect of an armrest on neuromuscular admittance is a novel finding.

For lower frequencies, the presence of the armrest decreases the B2P magnitude for all three tasks. Below frequencies of approximately 2 Hz, the ratio function is the lowest for the RT, signifying the largest decrease in B2P magnitude. For this task, the ratio function's magnitude, $|RF_{B2P}|$, is around 0.3, meaning that the B2P magnitude measured with armrest is reduced to 30% of the magnitude measured without armrest. For higher frequencies, it is the PT that shows the largest decrease in B2P magnitude, showing an $|RF_{B2P}|$ between 0.3 and 0.4. For high frequencies, above 10 Hz, the ratio function is close to $(1 + 0j)$, indicating no observable influence of an armrest on the B2P dynamics at higher frequencies. An interesting feature can be observed around 5 Hz. At this frequency, peaks occur in the ratio function for each control task. Strikingly, the PT shows a *reduction* in B2P magnitude ($|RF_{B2P}| \approx 0.3$) while the RT and FT show an *increase* in B2P magnitude ($|RF_{B2P}| \approx$ 2-3). The cause of these peaks is yet unknown.

A possibly related observation was made by Schoenberger and Wilburn [1973], who investigated tracking task performance under different vibration conditions using a side-stick and an armrest and found a large performance decrement for motion disturbances of 10 Hz. The authors attributed this to an increase in the transmission of vibrations to the subject's hand via the armrest at this frequency. The increase in BDFT effects at 5 Hz found in the current study may be related to the earlier finding. Possible explanations for the difference in frequency between the two observations may be differences in the armrest dynamics, armrest location and subject's posture.

It should be noted that the B2P magnitude around 5 Hz frequency is relatively small, reducing the importance of this feature within the overall B2P dynamics, at least in the open-loop case. In the case of a closed-loop BDFT system, such a peak could lead to poorly damped or unstable oscillations (especially in a vehicle with a structural mode around the same frequency). By attaining a stiffer neuromuscular setting, these problems can be reduced.

In Chapter 3 it was shown that:

$$H_{B2P}(s) = H_{FDFT}(s) \left(H_{CDFT}(s) + H_{B2FOL}(s) \right) \qquad \boxed{7.12}$$

As we already saw that the FDFT dynamics are not strongly affected by the addition of an armrest, and we know that the CDFT dynamics are constant, the changes in the B2P dynamics that are caused by the addition of an armrest need to be largely due to changes B2FOL dynamics. Recall that the biodynamic feedthrough to forces in open-loop (B2FOL) dynamics represent an acceleration-force coupling, i.e., these dynamics determine how acceleration disturbances result into involuntary forces that are applied to the control device. The relationship between B2P, B2FCL and B2FOL is illustrated in Fig. 7.7, which is repeated here from Chapter 3.

The results shown in Fig. 7.8 provide an insight in how the acceleration-force coupling is influenced by the addition of an armrest. The B2FOL dynamics that are shown were corrected for the influence of CDFT dynamics. The results show that the addition of an armrest has similar effects on B2FOL dynamics as on the B2P dynamics. For

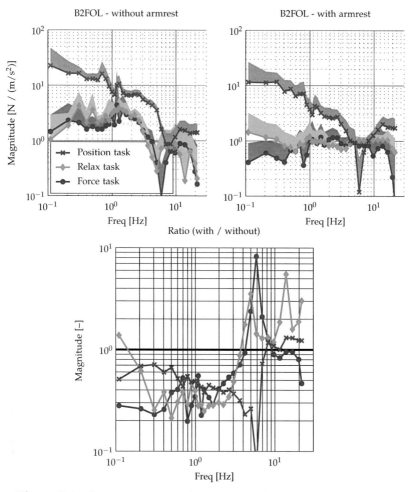

Figure 7.8: The B2FOL magnitude estimate, without armrest (left), with armrest (right) and the ratio function (bottom). The lines show means, the colored bands show standard deviation (mean + SD shown).

low frequencies the occurrence of involuntary forces is attenuated[a]. Around 5 Hz peaks occur, indicating an increase in feedthrough for the RT and FT, and a decrease for the PT. Note that the magnitude of these peaks is larger than for the B2P dynamics.

There are two main conclusions that can be drawn at this point. First of all, it can be said that the armrest is an effective tool in mitigating BDFT, as the addition of an armrest strongly reduces BDFT effects at low frequencies (< 2 Hz). As BDFT effects at these frequencies are likely to interfere with voluntary control inputs, the addition of an armrest would increase control performance in situations that otherwise would be vulnerable to BDFT. Furthermore, it can be concluded that the armrest used in this study did not influence the force disturbance responses (admittance and FDFT) substantially. Hence, the changes in B2P dynamics are largely due to changes in the B2FOL dynamics. Hence, it can be concluded that the way in which an armrest reduces BDFT is by reducing the amount of involuntary forces that are applied to the control device. One type of BDFT dynamics is not addressed yet: the biodynamic feedthrough to forces in closed-loop (B2FCL) dynamics. Recall that the B2FCL dynamics are related to the B2P dynamics through the known and invariant control device dynamics (Chapter 3):

$$H_{B2P}(s) = H_{CD}(s)\left(H_{CDFT}(s) + H_{B2FCL}(s)\right) \qquad \boxed{7.13}$$

where H_{CD} are the control device dynamics and the H_{CDFT} are the control device feedthrough dynamics As a consequence, the information contained in the B2FCL dynamics is not fundamentally different from the information contained in the B2P dynamics.

A complication with B2FCL is that its output, $F_{arm}(t)$, is enclosed in a closed-loop system and contains contributions from both the $M_{dist}(t)$ and the $\theta_{cd}(t)$ signals, as can be seen in Fig. 7.7. This fact makes B2FCL difficult to interpret.

Fig. 7.9 shows the results for the B2FCL dynamics. The results were corrected for CDFT dynamics. Clearly, the addition of the armrest

---

[a]The apparent increase in B2FOL for the RT for the lowest frequency seems an unlikely result and is possibly caused by the inaccuracies arising from obtaining the B2FOL dynamics by interpolation of the FDFT dynamics. See Chapter 3 for details on obtaining the B2FOL dynamics.

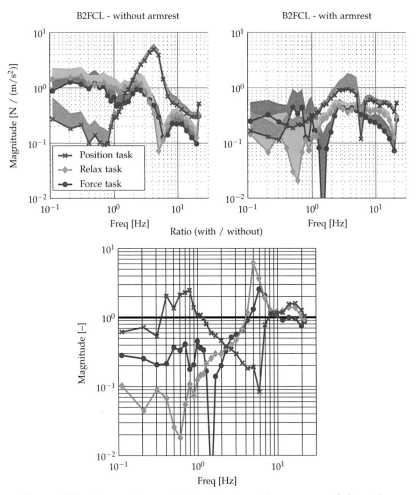

Figure 7.9: The B2FCL magnitude estimate, without armrest (left), with armrest (right) and the ratio function (bottom). The lines show means, the colored bands show standard deviation (mean + SD shown).

has a large influence on the dynamics. Interestingly, the ratio function shows that the B2FCL dynamics are influenced in a very different way by the addition of the armrest than the B2P dynamics were. For the low frequencies, the addition of an armrest strongly reduces the B2FCL dynamics for the RT and FT, even stronger than was observed for the B2P dynamics. The PT dynamics are less strongly attenuated and even show an increase in magnitude ($|RF| > 1$) between approximately 0.4 Hz and 1 Hz. At 5 Hz peaks in the ratio function can be observed. These are similar to the ones observed for the B2P dynamics in direction (i.e., a reduction for PT and increase for RT and FT) but have larger magnitudes.

The observation that the addition of the armrest leads to an increase in B2FCL magnitude, which implies an increase in $F_{arm}(t)$, for a PT between 0.4 Hz and 1 Hz seems mysterious, as we observed earlier that the addition of an armrest reduced both B2FOL and B2P dynamics for those frequencies and this task. Explaining this result requires unraveling the closed-loop dynamics.

Let's take a closer look at the changes that occur when an armrest is added, focusing our attention on the PT and disturbance frequencies between 0.4 Hz and 1 Hz. As shown in Fig. 7.7, the disturbance signal $M_{dist}(t)$ yields a force $F_{b2fol}(t)$ through the B2FOL dynamics. The result obtained for the B2FOL dynamics, shown in Fig. 7.8, showed that this force decreases when an armrest is added. This force is summed with $F_{nms}(t)$ and $F_{cdft}(t)$ to form $F_{app}(t)$, which causes the control device to deflect to angle $\theta_{cd}(t)$. The B2P results in Fig. 7.6 showed that this deflection also decreases when an armrest is added. The explanation why the B2FCL dynamics, and thus the $F_{arm}(t)$ force, increase lies in how the $\theta_{cd}(t)$ signal results in a $F_{nms}(t)$ force. We observed in the admittance results in Fig. 7.4 that the addition of the armrest caused a decrease in neuromuscular admittance for the PT below 1 Hz. A *decrease* in admittance implies stiffer neuromuscular dynamics, for which a deflection of the control device $\theta_{cd}(t)$ result in an *increase* in the force $F_{nms}(t)$. The B2FCL results show that, for the PT between 0.4 Hz and 1 Hz, the increase in $F_{nms}(t)$ is larger than the decrease in $F_{b2fol}(t)$, which results in an increase in B2FCL dynamics.

The main reason for going into some length to explain a feature of the B2FCL results is to illustrate that the closed-loop nature of the B2FCL dynamics makes it difficult to interpret results. Where the results of the FDFT, B2FOL and B2P dynamics allow for a fairly direct interpretation, the results of the B2FCL dynamics require a lengthy and finally somewhat unsatisfactory explanation. Hence, when studying the effect of an armrest (or any other mitigation approach) it is advisable to focus mainly on B2P and B2FOL dynamics, as these provide a clearer insight into the effects.

One may wonder why the B2FCL dynamics are so difficult to handle. After all, Eq. 7.13 showed that the dynamics are related to the B2P dynamics through the invariant control device dynamics. It can be easily demonstrated that the B2FCL ratio function is indeed related to the B2P ratio function through these dynamics, by combining Eq. 7.6 and Eq. 7.13:

$$RF_{B2P}(j\omega_m) = \frac{H_{CD}(j\omega_m)\left(H_{CDFT}(j\omega_m) + H_{B2FCL}^{sup}(j\omega_m)\right)}{H_{CD}(j\omega_m)\left(H_{CDFT}(j\omega_m) + H_{B2FCL}^{nosup}(j\omega_m)\right)} \qquad \boxed{7.14}$$

which can be simplified to:

$$RF_{B2P}(j\omega_m) = RF_{B2FCL}^{+}(j\omega_m) = \frac{H_{CDFT}(j\omega_m) + H_{B2FCL}^{sup}(j\omega_m)}{H_{CDFT}(j\omega_m) + H_{B2FCL}^{nosup}(j\omega_m)} \qquad \boxed{7.15}$$

This result shows that the ratio function of the B2P dynamics RF_{B2P} differs from the ratio function of the B2FCL dynamics RF_{B2FCL} only in the addition of the CDFT dynamics to both the numerator and denominator. Recall from Chapter 3 that the sum of the CDFT dynamics and the B2FCL dynamics are the 'uncorrected' B2FCL dynamics, where the CDFT dynamics are lumped into the B2FCL dynamics. The uncorrected B2FCL dynamics are noted as B2FCL^{+}, the ratio function of uncorrected B2FCL dynamics is noted as RF_{B2FCL}^{+}. Eq. 7.15 implies that the ratio function of the B2P dynamics RF_{B2P} is equal to the ratio function of the uncorrected B2FCL dynamics, RF_{B2FCL}^{+}. Fig. 7.10 shows the comparison between RF_{B2P}, obtained through Eq. 7.6, and RF_{B2FCL}^{+}, obtained through Eq. 7.15. The results are indeed identical.

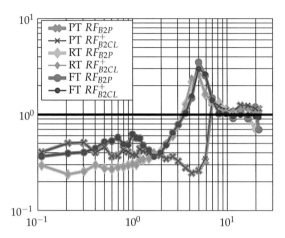

Figure 7.10: The ratio function of the B2P dynamics and the ratio function of the uncorrected B2FCL$^+$ dynamics.

The conclusion that can be drawn from the above analysis of the B2FCL dynamics is that it is not very suitable for use when studying the effect of a BDFT mitigation approach. This in contrast to BDFT modeling, where the B2FCL dynamics have proven to be very useful, as was demonstrated in Chapter 5. The reason the B2FCL dynamics fail to be of use when evaluating a BDFT mitigation approach is mainly due to its closed-loop character, which complicates its interpretation. This is not the case with the other dynamics that were analyzed, i.e., the admittance and FDFT dynamics and the B2P and B2FOL dynamics. In addition, the correction of the B2FCL dynamics for CDFT dynamics has a strong influence on its dynamics and ratio function, which further obscures its relationship to, e.g., the B2P dynamics.

7.5   Conclusions

The effectiveness of an armrest in mitigating biodynamic feedthrough was investigated. The results show that, generally, the presence of an armrest decreases the level of both B2P and B2FOL. That is, the involuntary deflections of the control device and involuntary forces applied to the control device are reduced by the addition of an armrest. This holds for each of the three levels of neuromuscular admittance investigated. The results furthermore provide the novel insight that the effect of the armrest varies strongly with frequency and neuromuscular admittance. Only a minor effect was observed of the addition of the armrest on the force disturbance responses, i.e., the admittance and FDFT dynamics, which can be explained by the location of the armrest. This observation supports the conclusion that all differences in the BDFT dynamics are due to the armrest directly and not due to changes in the neuromuscular admittance.

The results presented in this chapter show that an armrest is an effective tool in mitigating BDFT. The reduction obtained, especially at low frequencies, may very well be sufficient to obtain adequate task performance and prevent closed-loop oscillations in many practical situations which are currently suffering from the occurrence of BDFT. The fact that an armrest is cheap to produce and install makes this simple hardware component a viable alternative to more advanced mitigation methods.

The analysis also showed that the B2FCL dynamics are not very suitable for use when studying the effect of a mitigation approach. Due to its closed-loop character it lacks a clear interpretation. This is not the case for the other dynamics that were analyzed, i.e., the admittance, FDFT, and B2P and B2FOL dynamics.

More investigation is required to better understand the observed effects. For example, the peaks observed around 5 Hz remain unexplained but may be related to an observation made in [Schoenberger and Wilburn, 1973]. Also, the effect of the location and layout of the armrest was not investigated here. It is likely that changes in the dynamical properties of the armrest, such as the damping, may improve or deteriorate its effectiveness in BDFT mitigation. It was hypothesized that an armrest located closer to the wrist may have a stronger influence on the force disturbance responses (admittance and FDFT dynamics). Such an influence would have indirect repercussions on the BDFT dynamics as well, as these strongly depend on the admittance. A study investigating the effect of armrest parameters on BDFT dynamics, using techniques proposed in this chapter, would not only yield new insights with respect to BDFT mitigation, but also regarding the interrelation between admittance and BDFT dynamics.

CHAPTER 8

Admittance-adaptive model-based biodynamic feedthrough cancellation

This chapter presents a novel approach to BDFT mitigation. What differentiates this approach from others is that it accounts for adaptations in the neuromuscular dynamics of the human body. The approach was tested, as proof-of-concept, in an experiment where participants inside a motion simulator were asked to fly a simulated vehicle through a virtual tunnel. By evaluating the tracking performance and control effort with and without motion disturbance active and with and without cancellation active, the cancellation approach was evaluated. Results showed that the cancellation approach was successful and largely removed the negative effects of BDFT on performance and effort.

The contents of this chapter are based on:

Paper title	Admittance-adaptive Model-based Cancellation of Biodynamic Feedthrough
Authors	Joost Venrooij, Mark Mulder, David A. Abbink, Marinus M. van Paassen, Max Mulder, Frans C. T. van der Helm, and Heinrich H. Bülthoff
Published at	2014 IEEE International Conference on Systems, Man and Cybernetics, October 5-8, 2014, San Diego, USA

- and -

Paper title	Admittance-adaptive Model-based Approach to Mitigate Biodynamic Feedthrough
Authors	Joost Venrooij, Max Mulder, Mark Mulder, David A. Abbink, Marinus M. van Paassen, Frans C. T. van der Helm, and Heinrich H. Bülthoff
Submitted to	IEEE Transactions on Cybernetics

8.1   Introduction

A main motivator for previous and current BDFT research is the desire to *reduce* the BDFT effects on manual control performance. Several mitigation techniques have been suggested in the literature. For instance, an Active Vibration Isolation System (AVIS) was developed in [Schubert et al., 1970] and tested in [Dimasi et al., 1972], which isolates the human operator from vehicle accelerations by actively compensating for platform accelerations. A similar approach is used in recent work on backhoe excavators [Humphreys et al., 2014]. An adaptive filtering technique was proposed in [Velger et al., 1984] and tested in [Velger et al., 1988]. Results of yet another approach, called force reflection, were presented in [Repperger, 1995]. Force reflection relies on opposing the involuntary force caused by BDFT effects and so canceling its effects. This technique was also successfully used in [Sövényi, 2005]. In [Sirouspour and Salcudean, 2003] a robust controller to suppress BDFT effects using μ-synthesis was developed. These examples illustrate the fact that a range of studies have been devoted to BDFT mitigation and that many different ways of achieving that goal exist. For a review of the possible BDFT mitigation methods, the reader is referred to Chapter 6.

The current chapter aims to contribute to the existing body of knowledge regarding BDFT mitigation, by proposing and experimentally evaluating a novel mitigation approach: an admittance-adaptive, model-based signal cancellation technique. The most important contribution of this technique is the inclusion of an important influence on BDFT dynamics that has thus far not, or at least not systematically, been accounted for in other mitigation schemes: the adaptive neuromuscular dynamics of the human body. The human body dynamics vary between persons, for example due to different body sizes and weights, and also within one person over time. Humans can adapt their body's neuromuscular dynamics through muscle co-contraction and modulation of reflexive activity in response to, e.g., task instruction, workload and fatigue [Abbink and Mulder, 2010; Mulder et al., 2011]. The results in previous chapters have shown that the highly variable human body dynamics are amongst the most influential, complex and poorly understood factors in the large spectrum of properties determining the occurrence of BDFT. For a successful mitigation of BDFT, these variabilities must be understood and accounted for [Venrooij et al., 2011b].

The mitigation approach proposed in this chapter is *model-based*: it relies on a model to 'predict' the involuntary BDFT-induced control inputs. Mitigation is done using *signal cancellation*, that is, by subtracting the predicted involuntary input from the total input – which contains both voluntary and involuntary components – BDFT is canceled. This in contrast to the alternative technique of force cancellation, where the involuntary force is reflected by the control device. Finally, the approach is *admittance-adaptive*, a term used here to indicate the model's ability to account for changes in the settings of the neuromuscular system. The dynamics of the neuromuscular system are commonly described through the neuromuscular admittance [Abbink et al., 2011].

The novel admittance-adaptive feature of the proposed BDFT mitigation approach will be implemented in an elementary fashion, as a proof-of-concept. The approach was tested in an experimental setup where participants inside a motion simulator were asked to fly a simulated vehicle through a virtual tunnel, a so-called 'highway-in-the-sky' (HITS). Using measurements with and without motion disturbances active, and with and without cancellation active, the method was evaluated.

The chapter is structured as follows: first, the BDFT system model is briefly revisited in Section 8.2. Some general considerations regarding BDFT mitigation in are provided in Section 8.3. Based on these, the mitigation approach is discussed in Section 8.4, followed by a detailed experiment description in Section 8.5. The results are presented in Section 8.6, followed by the conclusions in Section 8.7.

8.2   Biodynamic feedthrough system model

The BDFT system model, shown in Fig. 8.1, helps to gain an understanding of the elements that play a role in biodynamic feedthrough. It contains all the high-level elements of a general BDFT

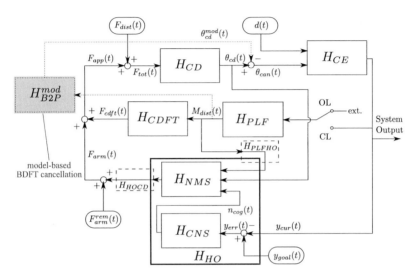

Figure 8.1: The biodynamic feedthrough system model. A human operator (HO) controls a controlled element (CE) using a control device (CD). Motion disturbances $M_{dist}(t)$ are coming from the platform (PLF). The feedthrough of $M_{dist}(t)$ to involuntary applied forces $F_{arm}(t)$ and involuntary control device deflections $\theta_{cd}(t)$ is called biodynamic feedthrough (BDFT). The feedthrough of $M_{dist}(t)$ to inertia forces $F_{cdft}(t)$ is called control device feedthrough (CDFT). $F_{app}(t)$ is the sum of the forces applied to the control device by the HO. The HO consists of a central nervous system (CNS) and a neuromuscular system (NMS). The connection between the HO and the environment is governed by two 'interfaces', H_{PLFHO} and H_{HOCD}. The CE and PLF can form an open-loop (OL) or closed-loop (CL) system. The focus of the current chapter: an adaptive-model model-based signal cancellation approach is developed, in which a BDFT model H_{B2P}^{mod} estimates the involuntary part of the CD deflections, $\theta_{cd}(t)$. The model output is subtracted from the total control device deflections to cancel BDFT.

system. For an elaborate discussion of the BDFT system model, the reader is referred to Chapter 3.

Recall that due to the two disturbance inputs, motion disturbance $M_{dist}(t)$ and force disturbance $F_{dist}(t)$, the control device deflection signal $\theta_{cd}(t)$ consists of a number of contributions:

$$\theta_{cd}(t) = \theta_{cd}^{Fdist}(t) + \theta_{cd}^{Mdist}(t) + \theta_{cd}^{cog}(t) + \theta_{cd}^{rem}(t) \qquad \boxed{8.1}$$

where superscripts ^{Fdist} and ^{Mdist} mark the contributions of the force and motion disturbances. Note that $\theta_{cd}^{Mdist}(t)$ is the involuntary part of the control input caused by the acceleration signal, i.e., the BDFT effect. The superscript ^{cog} denotes the cognitive part in the CD deflection, i.e., the deflection due to voluntary control actions coming from H_{CNS}. The remaining part, the remnant, is denoted with the superscript ^{rem}. It can be defined as the operator's control output that is not linearly correlated with the system inputs (here the disturbance signals) [McRuer and Jex, 1967]. In the BDFT system model, remnant originates from the remnant force signal $F_{arm}^{rem}(t)$.

Model-based signal cancellation is indicated in Fig. 8.1 with dotted lines: motion disturbance signal $M_{dist}(t)$ forms the input for the BDFT model H_{B2P}^{mod}. The model output, $\theta_{cd}^{mod}(t)$, represents an *estimate* of the involuntary part of control device deflections $\theta_{cd}^{Mdist}(t)$, and is subtracted from the total control device deflection signal $\theta_{cd}(t)$. The result is a 'corrected' control device deflection signal, $\theta_{can}(t)$, which enters the CE. If the contribution $\theta_{cd}^{Mdist}(t)$ is largely canceled by $\theta_{cd}^{mod}(t)$, the cancellation is successful. Note that through signal cancellation, the actual, physical deflection of the CD is *not* changed, only the input to the CE is adjusted. Also note that if cancellation is *not* active, $\theta_{can}(t)$ is equal to $\theta_{cd}(t)$.

8.3 Mitigation considerations

Having introduced the BDFT system model and the important BDFT elements, some considerations can be discussed regarding BDFT mitigation.

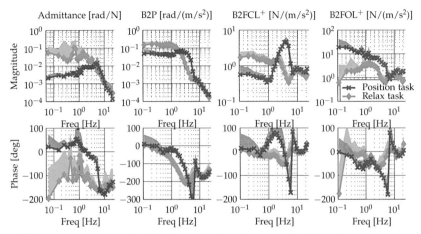

Figure 8.2: Example of admittance, biodynamic feedthrough to positions (B2P), BDFT to forces in closed-loop (B2FCL$^+$) and BDFT to forces open-loop (B2FOL$^+$) dynamics, for two different classical tasks, averaged over 12 subjects (see [Venrooij et al., 2011a]).

8.3.1 Between- and within-subject variability

Biodynamic feedthrough is known to depend on many different factors [McLeod and Griffin, 1989]. Using Fig. 8.1 they can be broadly identified. Some examples: the control device dynamics H_{CD} affect how involuntary forces result in involuntary deflections ([McLeod and Griffin, 1989; Venrooij et al., 2013a] and Chapter 3); adding or removing an armrest changes the H_{HOCD} dynamics and with that the BDFT dynamics ([Torle, 1965; Venrooij et al., 2012] and Chapter 7). There are many more factors which play a role, but when considering BDFT for one particular vehicle, with a certain cockpit layout, most of these influencing factors become invariant.

An important exception to this is the human operator. The variabilities between and within human operators renders BDFT a variable, dynamic relationship, both varying *between* different persons (between-subject variability), but also *within* one person over time (within-subject variability). Previous research (Chapter 6) suggests that for a successful model-based cancellation of BDFT, both between- and within-subject variability need to be taken into account, as otherwise the cancellation might fail.

What this implies: for successful cancellation, the BDFT model needs to be both *personalized* for each subject and *adapted* to the situation at hand. Section 8.3.3 addresses how this was done in the current study.

8.3.2   Types of biodynamic feedthrough

Before constructing a BDFT model, a choice needs to be made regarding which type of BDFT dynamics is to be modeled. It was proposed Chapter 3 that the BDFT response can be separated in several related dynamical relationships. The first distinction that can be made is that BDFT causes both involuntary *positions* (deflections) of the CD and involuntary *forces* applied to the CD. This notion leads to the concepts of *BDFT to positions* – abbreviated as B2P – and *BDFT to forces* – abbreviated as B2F.

The former, B2P, describes the transfer dynamics from vehicle accelerations (e.g., in $[m/s^2]$) to involuntary control device deflections (e.g., in [rad]). For the latter, B2F, two variations exist. One way of obtaining B2F is determining the dynamics between vehicle accelerations M_{dist} (in $[m/s^2]$) and forces $F_{arm}(t)$ (in [N]). These dynamics are labeled *B2F in closed loop* (B2FCL) as the force signal $F_{arm}(t)$ is contained in a closed-loop system. Another way of obtaining B2F is by 'opening' this loop and calculating the transfer dynamics between accelerations and the open-loop force applied to the CD (not shown in Fig. 8.1). This open-loop force is that part of the force signal $F_{arm}(t)$ which is directly and solely due to the acceleration disturbances (see Chapter 3 for details). These dynamics are called *B2F in open loop* (B2FOL).

An example of the different BDFT dynamics are shown in Fig. 8.2. Without going into the details regarding the differences and interrelations between these dynamics, it is important to note that all three types of BDFT dynamics are candidates for modeling. Choosing one of them determines the type of cancellation that can be performed.

Force cancellation requires a B2FOL model, in order to model and reflect the open-loop involuntary force. For signal cancellation – the type of cancellation applied here – a B2P model is required. There are several benefits in exploiting the relationship between B2P and B2FCL$^+$ dynamics in this case. It can be shown that:[a]

$$H_{B2P}(s) = H_{CD}(s)H_{B2FCL}^+(s) \qquad \boxed{8.2}$$

implying that the B2P dynamics can be easily obtained from the B2FCL$^+$ dynamics. A possible way of interpreting this is that the B2P dynamics are the result of filtering the B2FCL$^+$ dynamics through the CD dynamics. This is an important argument to select B2FCL$^+$ dynamics for modeling, as the important features are unfiltered and therefore more pronounced (compare B2P and B2FCL$^+$ dynamics in Fig. 8.2). So, by constructing the model on B2FCL$^+$ data and then multiplying the model with the known CD dynamics, a B2P model is obtained.

What this implies: for signal cancellation, a B2P model is required, which can be obtained through a B2FCL$^+$ model, multiplied with the known CD dynamics.

8.3.3   Neuromuscular admittance

Neuromuscular admittance represents limb dynamics by describing the relation between a force input and a position output of a limb. The admittance can vary strongly through muscle co-contraction and modulation of reflexive activity in response to, e.g., task instruction, workload and fatigue [Abbink and Mulder, 2010; Mulder et al., 2011]. As humans in actual vehicle control tasks also change their neuromuscular settings to the task at hand, these variations in BDFT dynamics need to be accounted for.

One way of varying neuromuscular admittance in a repeatable and controllable way is to use the so-called classical tasks [Abbink, 2006]: the position task (PT), the force task (FT) and the relax task (RT). For

---

[a]Note that this relationship makes use of so-called 'uncorrected' B2FCL dynamics, indicated with a superscripted $^+$. The dynamics are not corrected for CDFT dynamics. As the final goal is to create a B2P model, correcting the B2FCL dynamics is not required, see Chapter 3.

the PT the instruction is to keep the position of the control device in the centered position, that is, to "resist perturbations as much as possible". For the FT the instruction is to minimize the force applied to the control device, that is, to "yield to the perturbations as much as possible"; relax task (RT), in which the instruction is to relax the arms while holding the control device, that is, to "passively yield to the perturbations". For the PT the best performance is achieved by being very stiff (low admittance), the FT requires the operator to be very compliant (high admittance). The RT yields an admittance reflecting the passive dynamics of the neuromuscular system.

These task instructions cause the human operator to attain a particular neuromuscular setting, resulting in variations in the neuromuscular admittance. Fig. 8.2 shows the admittance measured for the PT and RT. The B2P, B2FCL$^+$ and B2FOL$^+$ dynamics are also shown, which were measured simultaneously. Clearly, the setting of the neuromuscular system influences not only the admittance, but all three types of BDFT dynamics as well. As humans in actual vehicle control tasks also change their neuromuscular settings to the task at hand, these variations in BDFT dynamics need to be accounted for.

What this implies: successful model-based BDFT cancellation requires the model to account for variability in the BDFT dynamics due to neuromuscular adaptations; this feature can be referred to as an *admittance-adaptive* capability. In this study – which is intended as proof-of-concept of an admittance-adaptive approach – two settings of the neuromuscular system will be investigated: the 'relaxed' setting, associated with the RT and the 'stiff' setting, associated with the PT.

8.3.4   The role of cognitive corrective inputs

Little is known about how *voluntary* cognitive control inputs affect the involuntary BDFT-induced control inputs. One generally assumes that a human operator is capable of correcting for at least parts of involuntary inputs. As human control capabilities are limited in bandwidth, it is safe to assume that these cognitive corrections are also limited up to a certain frequency, e.g., to 1 Hz [Allen et al., 1973]. How exactly an operator realizes cognitive BDFT corrections, e.g., whether these are based on visual or proprioceptive information, or how they depend on workload and task difficulty, is largely unknown and requires further investigations.

The effects of cognitive corrective inputs are not the focus of the current chapter, but nevertheless, their presence may have some implications for the cancellation approach proposed here. Especially for the measurements in which data are obtained to construct the BDFT cancellation models, the *identification measurements*, the presence of cognitive corrective inputs can be, in fact, detrimental. This issue will be dealt with in more detail in the next section, but to already provide an intuition to what this means: if the human operator invests cognitive control effort in correcting (a part of) the BDFT effects, these effects are thus removed from the BDFT measurements. As a consequence, any BDFT model based on these data will not model these and, when using this model in model-based BDFT cancellation, will not correct for them either. This means that the *same* amount of cognitive control effort will be required from the operator whether mitigation is active or not, as these effects will still have to be corrected manually. In other words: the harder the operator works in the identification measurements, the 'lazier' the BDFT model will become.

What this implies: The presence of cognitive corrective inputs requires special attention, especially in the identification measurements on which the BDFT models are based.

8.4 Mitigation approach

With the considerations of the previous section in mind, this is a suitable moment to address the mitigation approach adopted here. In doing so, some overlap with the discussion of the experimental design, which will be presented in Section 8.5, cannot be completely avoided. Still, it is insightful to present a general description of the mitigation approach here, separately, first. More details will be provided in Section 8.5.

8.4.1   Challenges and opportunities

The goal of this study is to show a proof-of-concept for an admittance-adaptive model-based BDFT cancellation approach. This poses several challenges and opportunities. One of the opportunities is that such a proof can be obtained for a general case, which provides freedom in the selection of the level of realism of the vehicle dynamics and vehicle control task. One of the challenges, on the other hand, is that a truly admittance-adaptive solution, capable of adapting online, is hard to obtain. Techniques for robust, accurate online estimation of the neuromuscular admittance are not widely available, and if existing at all, it is only in an early experimental stage. The development of a BDFT cancellation approach relying on the detection of neuromuscular adaptation through online techniques is therefore currently still out of reach. Also other potential ways of implementing online adaptation of the BDFT model, e.g., through measuring the grip force, are not straightforward to realize as the mapping between grip force and BDFT dynamics is not well understood.

Even though the techniques required to make this concept applicable in practice are still to be developed, it is worthwhile to show its potential. The following paragraphs will explain how the challenges were met and opportunities were exploited.

8.4.2   Highway-in-the-sky

This study is loosely based on a rotorcraft application, without burdening the experiment with the complexities of realistic helicopter flight simulation. The control devices where those used in typical helicopters. The dynamics of the simulated vehicle were highly simplified helicopter roll dynamics. Participants were asked to fly this vehicle through a highway-in-the-sky (HITS), i.e., a virtual tunnel [Mulder and Mulder, 2005].

The HITS representation was chosen for two reasons. First, it provides a close-to-realistic vehicle control task, where a vehicle moves through three-dimensional space. The tunnel image provides performance bounds and reference to current and future target positions [Mulder and Mulder, 2005], similar to many real-life vehicle control tasks.

Second, the HITS provides a means to *impose an adaptation* of the neuromuscular dynamics on the subject. By changing the size of the tunnel frames, the performance bounds can be altered [Mulder and Mulder, 2005]: in a narrow tunnel, pilots will need to increase co-contraction and reflexive activity to stay inside the tunnel, and vice versa. This change occurs largely intuitively, but by additionally instructing subjects to react to the changes in the tunnel frames, a robust way of changing NMS dynamics during the experiment was obtained.

8.4.3 Neuromuscular adaptation

The HITS consisted of square tunnel frames of two sizes and colors: wide white frames and narrow red frames. The instruction for both types of frames was to stay inside the tunnel. For the wide white tunnel this should be done with minimum control effort, largely ignoring the exact vehicle position within the tunnel and only steering when necessary to remain inside the tunnel. For the narrow red tunnel, the instruction was to follow the center of the tunnel frames as closely as possible, requiring maximum control effort.

With these task instructions, the 'optimal' NMS settings varied between 'passive relaxed' and 'active stiff'. These settings correspond to the RT and PT classical tasks. Hence, a HITS section with wide white tunnel frames is referred to as an 'RT section', one with narrow red frames as a 'PT section'.

8.4.4 Model development step 1: Identification measurements

In the first step, the *identification measurements*, BDFT dynamics were measured for each subject while performing classical relax and position tasks, a method used in many previous studies (e.g., Venrooij et al. [2011a]). These measurements formed the basis of two BDFT models: one representing BDFT dynamics in the RT sections, the other the BDFT dynamics in the PT sections. In a later stage these models served to 'predict' the involuntary control inputs (as shown in Fig. 8.1).

The success of the model-based cancellation technique hinges on the assumption that the BDFT model is an accurate representation of the actual BDFT dynamics of the human operator. Therefore, it is desirable to perform the identification measurements in exactly the same conditions as where the cancellation will be applied afterwards. This implies having the subjects fly through the HITS, with the RT and PT sections, just as they will do during the cancellation experiment. However, this poses a problem regarding the effect of cognitive corrective inputs, briefly touched upon earlier. By showing a HITS, the subject can and will cognitively correct for involuntary BDFT effects. This results in poor BDFT data, as some BDFT effects, for which the controller should account, are no longer present in the data. A model based on this data will be somewhat 'lazy', as the same cognitive effort is required from the HO with and without controller active. That is, all BDFT effects that were canceled manually in the identification measurements will still need to be canceled manually.

A possible way of dealing with this is to correct the measured data for the influence of cognitive control actions. However, as still very little is known about this interaction, it would require thorough research to develop a sound way of doing so. In this study a more direct approach was selected: by removing the HITS in the identification measurements, the role of cognitive corrective inputs was minimized. The subject was asked to perform an RT or PT, without having any visual reference of performance. This is possible because the classical tasks do not require visual feedback. In this fashion 'clean' BDFT measurements were obtained, without cognitive interference of the operator.

An evident weakness of this approach is that a subject may behave differently with and without HITS. It is possible that the task constraints posed by the HITS result in different neuromuscular dynamics than attained due to a verbal task instruction without HITS. To minimize the impact of this issue, the subjects received training with HITS before the identification measurements started. They were specifically instructed to attain a consistent neuromuscular setting in the RT and PT sections and memorize it to the best of their abilities. In the identification phase they were asked to reproduce the neuromuscular setting used during the RT and PT sections. During the remainder of the experiment they were encouraged to attain a consistent setting of their neuromuscular dynamics for the two tunnel sections.

The approach adopted here is imperfect and it may result in differences between the modeled and actual BDFT dynamics. However, as long as neither the techniques for online neuromuscular dynamics identification nor ways to correct for the cognitive corrective inputs exist, the current approach, imperfect as it may be, seems to be the best option available. Moreover, proving that successful cancellation is possible using this less-than-optimal approach will provide additional confidence that the concept itself is sound and can be improved as new techniques emerge.

8.4.5   Model development step 2: Parameter estimation

In the second step, the data of the identification measurements were used to develop two B2P models. The model structure and parameter estimation approach are described in detail in Chapter 5. In short, the approach uses a modeling technique referred to as 'asymptote modeling': by combining a number of base functions, with particular asymptotic characteristics, the measured B2FCL$^+$ dynamics can be accurately modeled in the frequency domain. Chapter 5 describes how this modeling approach yields a mathematical model with 16 parameters, and the following model structure:

$$H^{mod}_{B2FCL}(s) = K H^1_B(s) H^2_B(s) H^3_B(s) H^4_B(s) H^5_B(s) \qquad \boxed{8.3}$$

where s is the Laplace operator, K is a scaling gain and the H^k_B terms are the base functions (where k is the base function number). Each base function has the following structure:

$$H_B^k(s, f_{nk}, \zeta_k, \gamma_k) = \left(1 + 2\zeta_k/(2\pi f_{nk}s) + s^2/(2\pi f_{nk})^2\right)^{\gamma_k} \quad \boxed{8.4}$$

where f_{nk} is the natural frequency in Hz, ζ_k the damping factor $[-]$ and γ_k the order $[-]$. Note that if $\gamma_k = -1$ the base function describes typical mass-spring-damper (MSD) dynamics. The orders of the five base functions were chosen based on the slopes of different sections of the B2FCL$^+$ dynamics. Chapter 5 showed they can be chosen as:

$$\gamma_1 = -1, \quad \gamma_2 = +2, \quad \gamma_3 = -2, \quad \gamma_4 = +2 \text{ and } \gamma_5 = -1$$

The remaining parameters of each base function are obtained by fitting the model on the measured B2FCL$^+$ dynamics. Then, by multiplying the obtained model with the known control device dynamics yields the B2P model is obtained:

$$H_{B2P}^{mod}(s) = H_{CD}(s)H_{B2FCL}^{mod}(s) \quad \boxed{8.5}$$

As the control device dynamics are described with 3 parameters, the resulting B2P model contains 19 parameters.

8.4.6   Model development step 3: Implementation

In this study the adaptation of the model to the two neuromuscular settings, PT and RT, was implemented using a simple switching strategy based on the vehicle location in the HITS. When positioned in a PT section, the PT model was used (was 'active'); in an RT section, the RT model was used.

Note that, in this experiment, the subject only controlled the virtual vehicle and *not* the motion of the simulator. Hence, the acceleration disturbance signal is independent from the inputs provided by the operator, making this an 'open-loop' experiment which allowed us to accurately measure the BDFT dynamics also during the cancellation experiment (see also Section 8.6.7). In this study the input to the B2P model was the motion disturbance signal $M_{dist}(t)$. One could also obtain this signal from an Inertial Measurement Unit

(IMU), but this was not pursued in this proof-of-concept study. The input was corrected for the delay between sending out the motion command and the occurrence of a BDFT response. This delay was determined to be in the order of 60 ms (of which approximately 40 ms is the delay of the motion platform). The $M_{dist}(t)$ signal was delayed for this duration before being used as input to the model:

$$\theta_{cd}^{mod}(t) = H_{B2P}^{mod} M_{dist}(t - t_d) \qquad \boxed{8.6}$$

where t_d is the delay time of 60 ms. The output $\theta_{cd}^{mod}(t)$ is a prediction of the involuntary control device deflection caused by BDFT effects.

8.4.7   The cancellation experiment: conditions

The cancellation experiment had two independent variables: the task (TSK) and condition (COND). The two TSK levels were PT and RT. There were four COND levels: the condition in which the cancellation was active will be referred to as the cancellation condition (CAN). In addition, two baseline conditions were used: a static condition (STA), in which no acceleration disturbances were applied (simulator not moving), and a motion condition (MOT), in which acceleration disturbances were applied but cancellation was inactive. Comparing tracking performance between the STA, MOT and CAN conditions provides insight in how the acceleration disturbances cause the performance to deteriorate and how much of the original performance can be restored with cancellation.

To investigate the importance of within-subject variability and the effects of neuromuscular settings on cancellation, a fourth condition was added, in which the cancellation was done in an 'incongruent' fashion, i.e., the PT model was applied in the RT HITS sections and vice-versa. This condition will be referred to as the incongruent condition (INC). The two independent variables TSK (PT, RT) and COND (STA, MOT, CAN, INC) resulted in a 2×4 within-subjects repeated-measures design.

8.4.8   The cancellation experiment: metrics

To evaluate the cancellation approach a range of different metrics can be used. For the current chapter, a selection was made of three metrics that were deemed most insightful. They provide information regarding (i) the quality of the cancellation (ii) the tracking performance (iii) the control effort.

The quality of the cancellation was evaluated using a cancellation percentage, introduced in Chapter 6. The percentage gives an indication for how much of the involuntary control inputs were canceled.

As a measure for performance an error metric was used: the average absolute heading error. The metric reflected how well the subject was able to follow the heading of the HITS. A large average heading error relates to a low performance.

The reason for using a heading error instead of a position error is that (i) the latter was not directly observable by the subject (the center of the tunnel was not indicated on the screen) and (ii) minimizing position error is not in line with the optimal control strategy when moving along a curved path with preview. In the RT sections the instruction was to stay inside the HITS using minimal control effort, allowing a subject to use the full width of the tunnel. A position error would be inadequate to evaluate the performance of such a task. In the PT sections, where the instruction was to stay as close as possible to the center of the tunnel frames, a position error could be considered to be more appropriate. However, the optimal (and natural) human control strategy is to exploit the preview of the tunnel to 'cut the curves', reducing rotational velocities and accelerations by optimizing the vehicle's future path rather than its current instantaneous position. In the pilot phase of the experiment, several tests were conducted using virtual markers to indicate the exact center of the tunnel and the position of the vehicle on the screen. This led to considerable disagreement between the position of the center marker and the natural path the subject would usually take (the marker was reported to "take the turns too late"). This resulted in a situation where the subject would either ignore the markers or the tunnel. By removing the markers and using heading error as performance metric, subjects were allowed to express natural control behavior while adhering to the task instructions.

To evaluate the effort, the derivative of the control device deflection, i.e., steering speed, was computed. A high steering speed can be considered to be related to a high effort. Speed is preferred above position, as it is less sensitive to the low frequency steering inputs required to follow the curvature of the HITS. As a metric for effort the root-mean-square (RMS) of the steering speed was calculated. A high value relates to high effort.

8.5 Experiment description

8.5.1 Hypotheses

The experiment aimed at testing three hypotheses:

- the BDFT hypothesis: BDFT will occur in the motion condition (MOT) and will result in decreased tracking performance and increased control effort as compared to the static condition (STA).

- the cancellation hypothesis: with cancellation active (CAN), performance will improve, effort will decrease as compared to the condition without cancellation (MOT); performance and effort are partially restored to the values obtained in the static condition (STA).

- the incongruency hypothesis: incongruent cancellation (INC) leads to lower performance and higher effort than obtained with congruent cancellation (CAN).

8.5.2 Apparatus

The experiment was performed in TU Delft's SIMONA Research Simulator (SRS) [Stroosma et al., 2003]. The control devices were electrically-actuated helicopter controls (collective and cyclic) with adjustable dynamics. For this experiment only the cyclic roll axis

Table 8.1: Control device dynamical settings.

Axis	Inertia [Ns^2/deg]	Damping [Ns/deg]	Stiffness [N/deg]	Length [mm]
Cyclic roll	0.0162	0.0516	1.000	650

Table 8.2: Data of subjects (N = 11).

	Age [years]	Weight [kg]	Height [cm]	BMI [kg/m^2]
mean	26.0	77.9	183.6	23.1
st. dev.	3.9	11.9	8.6	2.9
Range	21-35	57-100	172-196	19.3-27.8

was used for lateral control inputs. The dynamics used for this axis are listed in Table 8.1; no non-linearities were simulated.

A helicopter seat was used, in which the subjects were strapped-in with a five-point safety belt. Visual information was displayed on a head-down display (15-in LCD, 1024×768 pixels, 60 Hz refresh rate) in front of the subject.

8.5.3   Subjects

Data were collected for eleven male subjects, Table 8.2. One subject was left-handed, all others were right-handed. Their body mass index (BMI) was calculated by dividing a person's weight (in [kg]) by height squared (in [m^2]), and is a measure of the total amount of body fat in adults [Maddan et al., 2008]. Values between 19 and 25 are considered to be normal (mesomorph) [Maddan et al., 2008]. Two subjects had higher values (endomorph).

8.5.4   Task instruction

Recall that the HITS consisted of tunnel frames of two sizes and colors: wide white frames (RT section) and narrow red frames (PT section). The instruction for both sections was to stay inside the tunnel. For the RT section this should be done with minimum control effort, while keeping the arm 'relaxed' and largely ignoring the

Table 8.3: Repetitions per condition.

Straight HITS			
	STA	MOT	CAN	INC
PT	3	3	3	×
RT	3	3	3	×
Curved HITS			
	STA	MOT	CAN	INC
PT	6	6	6	6
RT	6	6	6	6

exact vehicle position within the tunnel and only steering when necessary to stay inside the tunnel. For the PT section the instruction was to stay as close as possible to the center of the tunnel frames, while keeping the arm 'stiff', using maximum control effort.

8.5.5 Experiment execution

Subjects received the above task instructions before entering the simulator. Once installed in the simulator, the subjects received training. To familiarize the subject with the vehicle dynamics, the HITS and the tasks, the first training run was performed without motion in a straight HITS with several RT and PT sections. The second training was performed in the same HITS, but with the motion disturbance activated. In the third and fourth training run the subject was presented with a curved HITS. In these runs, subjects were instructed to attain a consistent and repeatable neuromuscular setting in the different sections and memorize this setting to the best of their abilities.

After training, the subject's BDFT dynamics were measured in the identification measurements. Here, no visual information was displayed to the subject. The subject performed, in random order, five RTs and five PTs. When these were completed, the subject received a break of approximately 10 minutes, during which the model parameters were estimated.

In the cancellation experiment, subjects were presented with both straight and curved HITS sections (more details will follow). For the straight HITS, with PT and RT sections, the conditions STA,

MOT and CAN were tested with 3 repetitions for each subject. For the curved HITS, also with both PT and RT sections, the STA, MOT, CAN and – in addition – the INC conditions were tested with 6 repetitions for each subject. Table 8.3 shows the number of repetitions per condition for each subject. The subject received a 10 minute break after completing one-third and two-thirds of the total number of repetitions.

8.5.6 Vehicle dynamics

The vehicle that the subjects controlled moved forward at a constant speed of 25 m/s ($\approx$ 50 knots). The vehicle dynamics represented highly-simplified helicopter roll dynamics. These simplified dynamics were used for two main reasons. First, they resulted in a direct coupling between control inputs and vehicle response, yielding a strong correspondence between inputs and performance, making performance a more reliable metric. Second, because of the simple vehicle dynamics, the experiment could be performed by non-expert pilots, which facilitated data collection.

The vehicle responded only to lateral control inputs (roll). As can be seen in Fig. 8.1, the input for the vehicle dynamics was the signal $\theta_{can}(t)$, which was either equal to the control device deflection $\theta_{cd}(t)$ (in the STA and MOT conditions), or the difference between $\theta_{cd}(t)$ and the output of the BDFT model, $\theta_{cd}^{mod}(t)$ (in the CAN and INC conditions). The vehicle's roll angle ϕ was directly coupled to the control input:

$$\phi(t) = K_\phi \theta_{can}(t)$$

8.7

where K_ϕ is the roll gain, set to 0.05.

Every time step, the vehicle heading $\psi(t)$ was updated through numeric integration of the roll angle:

$$\psi(t) = \psi(t - \Delta t) + K_\psi \phi(t)\Delta t$$

8.8

where K_ψ is the heading gain and Δt is the simulation time step (0.01 s). In this study the heading gain was set to 5.0.

Table 8.4: Curved HITS parameters.

	K_s [m]	f_1 [m^{-1}]	f_2 [m^{-1}]	p_1 [deg]	p_2 [deg]
CUR-PTRT	100	$8 \cdot 10^{-4}$	$4 \cdot 10^{-4}$	180	12
CUR-RTPT	100	$8 \cdot 10^{-4}$	$4 \cdot 10^{-4}$	-20	0

8.5.7 HITS configuration

The HITS tunnel frames were 25 m apart, such that at a speed of 25 m/s approximately one frame would pass every second. The width and height of the RT section frames were 25 m, for the PT sections this was 5 m.

Four different HITS were used during the experiment:

- STR-PT: A straight HITS consisting of a PT section.

- STR-RT: A straight HITS consisting of an RT section.

- CUR-PTRT: A curved HITS with a PT section, followed by an RT section.

- CUR-RTPT: A curved HITS with an RT section, followed by a PT section.

Each RT and PT section consisted of 50 frames. In addition, each tunnel started with a straight lead-in section of 5 frames. The CUR tunnels had an additional straight lead-out section of 5 frames. In total, the STR tunnels had a length of 55 frames (5+50), the CUR tunnels had a length of 110 frames (5+50+50+5). During the cancellation experiment, the different HITS were presented in random order. Each HITS was repeated three times for each subject and each condition, yielding the number of repetitions listed in Table 8.3. Note that the CUR-type HITS contained both a PT and RT section, so three repetitions of both CUR-types resulted in six repetitions of the PT and RT condition.

The HITS trajectory was defined in an x-z coordinate frame. The curved trajectory was constructed by summing two sinusoids with

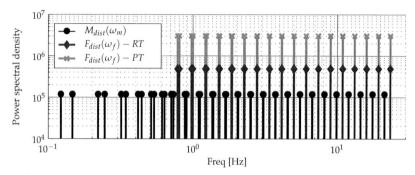

Figure 8.3: The power spectral density of the two disturbance signals M_{dist} and F_{dist}. The magnitude of F_{dist} was varied between the RT and PT sections.

a chosen frequency and phase shift:

$$x = K_s \sum_{k=1}^{2} \sin(2\pi f_k z + p_k \frac{\pi}{180}) \qquad \boxed{8.9}$$

where K_s is a scaling gain, z are values of the z-coordinate, and f_k and p_k are the frequency and phase shift of the sinusoid, respectively. The chosen parameters are listed in Table 8.4. The frequencies f_1 and f_2 were chosen low enough to not require any high frequency control inputs. With a speed of 25 m/s, the required steering frequency to follow the individual sinusoids were 0.02 Hz and 0.01 Hz, respectively, well below the frequency content of the two disturbance signals (discussed below). The difference between CUR-PTRT and CUR-RTPT, apart from the order of the PT and RT sections, was a phase shift. In the analysis, the results obtained in the CUR-PTRT and CUR-RTPT tunnels are combined.

8.5.8   Disturbance signals

In the cancellation experiment, two disturbance signals were applied simultaneously: a motion disturbance signal $M_{dist}(t)$ on the simulator's motion base and a force disturbance signal $F_{dist}(t)$ on the control device. The former signal was used to determine the BDFT (B2P and B2FCL[+]) dynamics, the latter signal was used to determine the neuromuscular admittance.

Both disturbance signals were multi-sines, defined by their frequency components. The length of the disturbance signals for the STR tunnels (with 55 frames) was 60 seconds, and for the longer CUR tunnels (with 110 frames) 120 seconds. These disturbance signal lengths were sufficient for the experiment.

The disturbance signals were separated in frequency, to allow distinguishing the response due to each disturbance in the measured signals [Abbink, 2006; Venrooij et al., 2011a]. The frequency content of the disturbance signals was equal in all conditions. For the motion disturbance signal $M_{dist}(t)$, 24 logarithmically-spaced pairs of frequency points, referred to as ω_m, were chosen between 0.1 and 21.4 Hz. The force disturbance $F_{dist}(t)$ was applied to measure admittance, a secondary objective. Hence, to minimize the influence of $F_{dist}(t)$ on control behavior and performance, the frequency range of $F_{dist}(t)$ was limited to high frequencies only: 18 logarithmically-spaced pairs of frequency points were chosen between 0.78 and 23.5 Hz; these frequency points are referred to as ω_f. There existed no overlap between ω_m and ω_f. Fig. 8.3 shows the power-spectral-densities (PSDs) of the two disturbance signals. The gain used for the $M_{dist}(t)$ signal was 0.8. For the disturbance signal used in the STR tunnel, the maximum acceleration was 3.67 m/s^2, maximum velocity was 1.12 m/s and maximum position was 0.72 m. For the 120 seconds disturbance signal (used in the CUR tunnel), these values were similar. For the force disturbance F_{dist}, the gain for the PT section was 5.0, and for the RT section 2.0 (note the difference in magnitude in Fig. 8.3). The gain was varied for two main reasons: (i) to keep the standard deviations of the control device deflections in the RT and PT sections approximately similar [Venrooij et al., 2011a], and (ii) as a 'haptic reminder' of the change in task, that is, the increase in force gain helped subjects to attain a stiffer neuromuscular setting in the PT sections. Gains were tuned to obtain high squared coherences for the admittance estimates [Venrooij et al., 2011a].

8.5.9 Calculating B2P and B2FCL

The applied force $F_{app}(t)$ and the control device deflection $\theta_{cd}(t)$ were measured, together with the vehicle state, i.e., its position, heading angle, roll angle, etc. In the analysis, signals $F_{dist}(t)$ and $M_{dist}(t)$ were used as commanded (not directly measured), and were cut to a length of 2^{12} samples (= 40.96 seconds). This allowed calculating the frequency response functions of the B2P and B2FCL dynamics. The truncated signals were also used to calculate three performance metrics.

B2P dynamics were estimated using the estimated cross-spectral density between $M_{dist}(t)$ and $\theta_{cd}(t)$ ($\hat{S}_{mdist,\theta}(j\omega_m)$) and the estimated auto-spectral density of $M_{dist}(t)$ ($\hat{S}_{mdist,mdist}(j\omega_m)$):

$$\hat{H}_{B2P}(j\omega_m) = \frac{\hat{S}_{mdist,\theta}(j\omega_m)}{\hat{S}_{mdist,mdist}(j\omega_m)}$$

8.10

where $\hat{H}_{B2P}(j\omega_m)$, with j the imaginary unit, is the estimate on frequencies ω_m of the B2P dynamics $H_{B2P}(s)$, with s the Laplace variable.

B2FCL$^+$ dynamics were calculated using the estimated cross-spectral density between $M_{dist}(t)$ and $F_{app}(t)$ ($\hat{S}_{mdist,f}(j\omega_m)$) and the estimated auto-spectral density of $M_{dist}(t)$ ($\hat{S}_{mdist,mdist}(j\omega_m)$):

$$\hat{H}^+_{B2FCL}(j\omega_m) = \frac{\hat{S}_{mdist,f}(j\omega_m)}{\hat{S}_{mdist,mdist}(j\omega_m)}$$

8.11

where $\hat{H}^+_{B2FCL}(j\omega_m)$ is the estimate of the actual B2FCL$^+$ dynamics $H^+_{B2FCL}(s)$ on frequencies ω_m. Note that by using the $F_{app}(t)$ signal, and not $F_{arm}(t)$, the $\hat{H}^+_{B2FCL}(j\omega_m)$ dynamics are not corrected for CDFT dynamics. As the dynamics are meant to be used to construct a B2P model (using Eq. 8.5) such a correction is not required. Neuromuscular admittance was estimated using $F_{dist}(t)$, but as these estimates will not be the subject of discussion here, the interested reader is referred to Chapter 3.

8.5.10   Performance metrics

Cancellation metric

A cancellation percentage was computed, introduced in Chapter 6, which indicates how much of the involuntary control inputs were canceled. This is evaluated by quantifying how much of the BDFT effects were removed when the model output $\theta_{B2P}^{mod}(t)$ was subtracted from the total control device deflection $\theta_{cd}(t)$. Using a frequency-decomposition technique Venrooij et al. [2011a], the total control device deflection $\theta_{cd}(t)$ can be decomposed in three parts:

$$\theta_{cd}(t) = \theta_{cd}^{Fdist}(t) + \theta_{cd}^{Mdist}(t) + \theta_{cd}^{res}(t) \qquad \boxed{8.12}$$

i.e., a contribution of F_{dist}, a contribution of M_{dist} and a residual, res. The residual is the sum of the cognitive and remnant contributions (see Eq. 8.1). The components of Eq. 8.12 were calculated by evaluating the PSD of $\theta_{cd}(t)$, on either ω_f, on ω_m or on all remaining frequencies. Through taking the inverse Fast Fourier Transform of these PSDs, the time series were obtained (see Chapter 2 for details on this operation). Applying the same operation, the components of $\theta_{can}(t)$ – the signal obtained after subtracting $\theta_{B2P}^{mod}(t)$ – could be obtained.

The cancellation percentage P_{can} was introduced in Chapter 6 as the ratio of the root-mean-square (RMS) of θ_{can}^{Mdist} and θ_{cd}^{Mdist}, which represent the M_{dist} component before and after the cancellation was applied:

$$P_{can} = \left(1 - \frac{RMS\left(\theta_{can}^{Mdist}(t)\right)}{RMS\left(\theta_{cd}^{Mdist}(t)\right)}\right) \cdot 100\% \qquad \boxed{8.13}$$

With a good B2P model, $\theta_{can}^{Mdist}(t)$ will be small compared to the original $\theta_{cd}^{Mdist}(t)$ and P_{can} will be close to 100%. In cases where the model is not as accurate, cancellation is less and P_{can} will be lower or even negative.

Error metric

Tracking performance was expressed using the average absolute heading error, μ_{ψ_e}. Heading error equals the difference between the current vehicle heading ψ and the target heading ψ_{tar}, defined as the heading of the section between the two frames where the vehicle is located:

$$\mu_{\psi_e} = \frac{1}{N} \sum_{k=1}^{N} |\psi_{tar}(k) - \psi(k)|$$

8.14

where N is the number of measurement samples. Note that a high value implies a low tracking performance.

The heading error shows how well subjects were able to follow the tunnel, independently from the actual position within the tunnel. In other words, subjects were not penalized for being left or right from the center (or, in fact, inside or outside the tunnel) but rather for not aligning the vehicle heading with the tunnel heading.

Control effort metric

As a measure for control effort, the RMS of the derivative of the control device deflections (the steering speed) was calculated. To improve the reliability of this measure as a metric for effort, the residual control device deflections $\theta_{cd}^{res}(t)$ were used (see Eq. 8.12). By using the residual component, instead of the complete control device deflection signal, the direct effects (i.e., the feedthrough) of the two disturbance signals were removed. This improves the reliability of the metric as direct feedthrough is not related to control effort. $\theta_{cd}^{res}(t)$ consists of the sum of the cognitive control inputs and remnant. Control effort was defined as:

$$E_{\dot{\theta}_{res}} = RMS(\dot{\theta}_{cd}^{res}(t))$$

8.15

where $\dot{\theta}_{cd}^{res}(t)$ is the time derivative of $\theta_{cd}^{res}(t)$.

8.6   Results

In the following, first the results for the curved HITS (CUR) will be presented in detail, followed by the results for the straight HITS (STR).

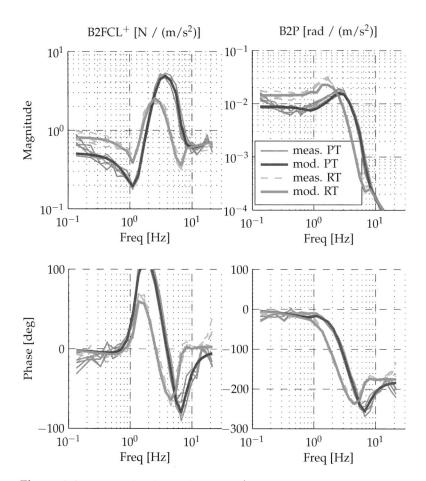

Figure 8.4: Measured and modeled B2FCL$^+$ and B2P dynamics (Subject #7). This subject showed large differences between the PT and RT conditions.

8.6.1   Identification measurements and parameter estimation

Results of the identification measurements for a selected subject (#7) are given in Fig. 8.4. The figure shows the measured B2FCL$^+$ and B2P dynamics for the PT and RT conditions. The five repetitions

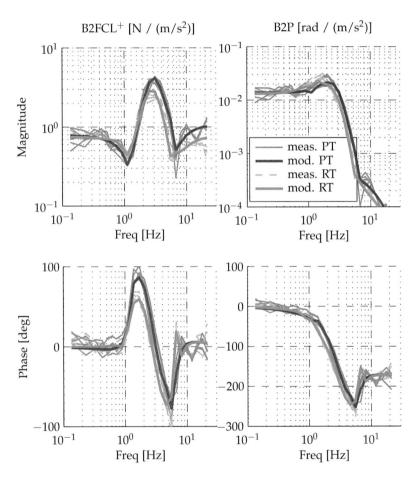

Figure 8.5: Measured and modeled B2FCL$^+$ and B2P dynamics (Subject #4). This subject showed small differences between the PT and RT conditions.

for each task resulted in very similar B2FCL$^+$ and B2P dynamics, which indicates consistency in the execution of the tasks. Clear differences in dynamics *between* the two tasks can be observed, illustrating the within-subject variability in BDFT dynamics. The dynamics of the PT and RT models, obtained by performing a parameter fit on the time average of the five repetitions, are also shown in Fig. 8.4. The models describe the measured dynamics well.

Not every subject showed such large differences between the two tasks. Fig. 8.5 shows results for another selected subject (#4). Here, for each task the repetitions yield similar results, indicating consistency in task execution. Model fits are still accurate, but differences between the RT and PT are small. It could be argued that this subject did not comply well to the task instruction, and did not vary the neuromuscular setting as instructed. Cancellation may still be successful, as long as the subject remains consistent throughout the experiment and uses the same neuromuscular settings during the cancellation experiment. If this is the case one would expect that for this subject similar levels of cancellation are observed in the CAN and INC case, as differences between the RT and PT models are small, which makes them interchangeable.

Comparing the results for these two subjects provides insight in the between-subject variability. The dynamics obtained for the PT show marked differences between the two subjects. This illustrates that an accurate description of the BDFT dynamics indeed requires *personalized* model parameters. The model parameters for each subject are listed in Table 8.5 on Page 319.

8.6.2 Cancellation metric

Results for the cancellation metric are given in Fig. 8.6, which shows the average (bars) and standard error of the mean (lines) of P_{can} (Eq. 8.13), obtained for the six repetitions of the PT and RT sections in the CAN and INC conditions for each subject. For both the PT and RT case, congruent cancellation (CAN) yielded a positive cancellation percentage in all conditions. For PT-CAN, the level of cancellation achieved varies between 33.5% and 61.0% and has an average (μ) of 47.0% and a standard deviation (σ) of 7.5%.

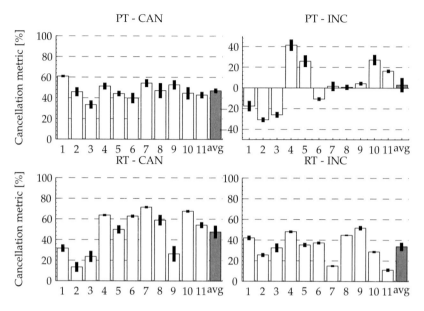

Figure 8.6: Percentage of BDFT cancellation for each subject (11 subjects) and the average (avg), for the PT and RT sections (row) and the CAN and INC conditions (column), curved HITS. The bars indicated the mean, the lines indicate the standard error of the mean.

For RT-CAN, the average is similar ($\mu = 47.6\%$) but the spread is much larger and cancellation varies between 13.4% and 71.6% ($\sigma = 20.2\%$). It can be concluded that for the PT a fairly robust result was obtained; cancellation was successful for each subject. For the RT, results vary much more between subjects.

In the INC condition, it can be seen that for both PT and RT tasks the cancellation reduces as compared to the CAN condition. Analysis of variance (repeated measures) showed a significant reduction in cancellation percentage from CAN to INC ($F(1,10) = 59.4$, $p < 0.001$) and from PT to RT ($F(1,10) = 29.3$, $p < 0.001$). Furthermore, a significant interaction between condition and task was found ($F(1,10) = 5.51$, $p < 0.05$), caused by the fact that the INC condition shows a much larger effect on the PT task than on the RT task. For PT-INC ($\mu = 3.0\%$, $\sigma = 23.2\%$), the cancellation percentage for some subjects became negative, meaning that incongruent cancellation decreased performance.

Subject #4 showed a relatively high level of cancellation for the INC condition (although still not as high as in the CAN condition). This was to be expected as this subject showed very similar behavior in the identification measurement (Fig. 8.5). The fact that still a relatively good cancellation is obtained in both the CAN and INC case, confirms that the subject attained consistent and similar neuromuscular settings throughout the experiment for both tasks.

Overall, the RT-INC condition has lower cancellation levels (μ = 33.9%, σ = 13.0%) than the RT-CAN condition. However, for four subjects the cancellation is higher: subjects #1, #2, #3 and #9 (note that these are also the subjects with the lowest RT-CAN results). There are two main ways of explaining this: (i) the RT model itself was of bad quality, and using the PT model instead yields better cancellation, (ii) subjects behaved differently in the identification measurements and the cancellation experiment.

Comparing the measured dynamics for the identification measurements with the model fit did not show any anomalies: the models showed adequate fits, comparable to those shown in Figs. 8.4 and 8.5. Hence, the RT model for these subjects was not of poorer quality than for other subjects, rejecting the first of the two explanations proposed above.

When looking at the dynamics measured while the model was actually applied (the CAN condition), a clear mismatch between modeled and actual behavior was observed for the subjects where cancellation results were poor. Fig. 8.7 shows the measured B2FCL$^+$ dynamics (magnitude only), obtained for the CAN condition, and the subject's model dynamics (i.e., the expected dynamics) for Subject #1. Note that the PT model matches well with the actual measured PT dynamics (good cancellation was also obtained for the PT-CAN condition). For the RT case, however, the model *over*estimates at low frequencies (<1.5 Hz) and *under*estimates at higher frequencies. At lower frequencies, the measured dynamics are little different from those measured for the PT. Hence, in the cancellation experiment this subject behaved 'stiffer', i.e., closer to the PT task,

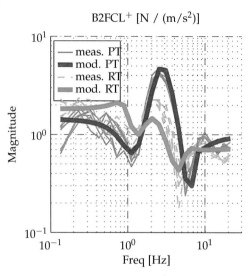

Figure 8.7: Measured and modeled B2FCL$^+$ dynamics in the CAN condition for Subject #1. This subject shows a mismatch between modeled and actual RT dynamics due to different behavior between the identification phase and the cancellation experiment.

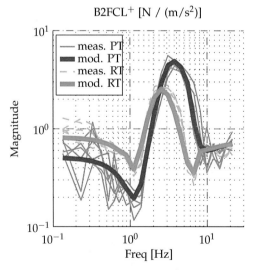

Figure 8.8: Measured and modeled B2FCL$^+$ dynamics in the CAN condition for Subject #7. This subject showed very similar behavior between the identification phase and the cancellation experiment.

such that the PT model used in the RT-INC condition yielded a *better* cancellation.

For comparison, results for Subject #7 are shown in Fig. 8.8. Here, the obtained PT and RT models provided an accurate description of the BDFT dynamics in the cancellation experiment, yielding high cancellation levels in the CAN condition and low levels in the INC condition. Also for other subjects the RT model provided a good description of the actual dynamics and good cancellation results were obtained in the CAN condition.

The above leads us to conclude that, overall, the cancellation was successful. Four subjects expressed different RT behavior between the identification measurements and the cancellation experiment, which led to the unexpected result that cancellation results in the RT-INC condition were better than in the RT-CAN condition. This issue was already highlighted as a possible weakness of the current approach, in which the BDFT dynamics were not monitored during the cancellation experiment. However, this should not be considered as a discouraging result. In fact, the observation that the subjects that show poor cancellation results in the RT-CAN condition show good results in the RT-INC condition further strengthens the conclusion that a model that better matches with the *actual* BDFT dynamics provides better cancellation results.

8.6.3 Tracking error metric

Figure 8.9 shows the error metric (averaged absolute heading error, where high values mean low performance) for all conditions: STA, MOT, CAN and INC. In this figure and the following, dashed gray lines connect the mean values obtained for each subject; bars show the mean across all subjects, with the black lines the standard error of the mean. Analysis of variance (repeated measures) showed a significant difference between conditions ($F(3,30) = 141.5$, $p < 0.001$) and tasks ($F(1,10) = 25.2$, $p < 0.001$) and a significant interaction between conditions and tasks ($F(3,30) = 26.9$, $p < 0.001$).

Comparing the results for the STA and MOT conditions, for each subject the heading error increased with motion present. That is: BDFT effects deteriorated tracking performance considerably. A

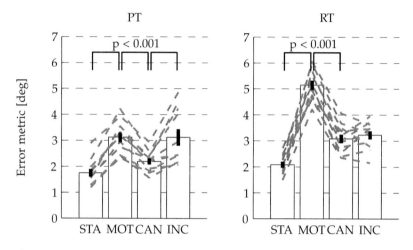

Figure 8.9: Tracking performance for the four conditions (PT and RT sections), curved HITS. The bars indicated the mean, the lines indicate the standard error of the mean. The dashed gray lines connect the mean values obtained for each subject.

contrast on this difference was calculated for both the PT and RT conditions and was significant (PT: $t(30) = 8.01$, $p < 0.001$ and RT: $t(30) = 20.30$, $p < 0.001$).

Results obtained for the CAN condition indicate that, for all subjects, performance *improved* due to the cancellation. For many subjects the error metric approximates the value obtained in the STA condition, signifying considerable benefit from the cancellation. Contrasts between the MOT and CAN conditions were significant, for both the PT and RT sections (PT: $t(30) = 5.42$, $p < 0.001$ and RT: $t(30) = 13.71$, $p < 0.001$). This confirms that the cancellation was indeed successful.

The importance of having an admittance-adaptive approach, which matches the cancellation to the current neuromuscular dynamics of the operator, is signified by the fact that for the PT the INC condition yielded an increase in the error metric, i.e., a deteriorated performance, compared to the CAN condition. Here, the contrast between the CAN and INC conditions was significant ($t(30) = 5.42$,

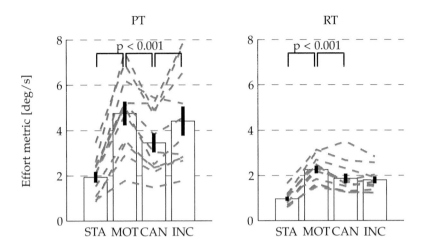

Figure 8.10: Control effort for the four conditions (PT and RT sections), curved HITS. The bars indicated the mean, the lines indicate the standard error of the mean. The dashed gray lines connect the mean values obtained for each subject.

$p < 0.001$). For the RT, the differences between CAN and INC are smaller, and in some cases the INC even shows a slightly better performance (Subjects #1, #2 and #9). These are the same subjects which showed better cancellation levels for the RT-INC than the RT-CAN condition, Fig. 8.6. For the RT, the contrast between the CAN and INC conditions was not significant ($t(30) = 0.951$, $p = 0.174$).

In conclusion, the performance data confirm the three experimental hypotheses. Performance deteriorates due to BDFT but can be largely restored when cancellation is active. In the PT, subjects performed better with congruent cancellation than with incongruent cancellation.

8.6.4   Effort metric

Figure 8.10 shows the effort metric (average RMS of the control device deflection derivative) for the four conditions: STA, MOT, CAN and INC. Here a high value indicates a high effort. Analysis of variance (repeated measures) showed a significant difference between conditions ($F(3,30) = 35.7$ $p < 0.001$) and tasks ($F(1,10) = 34.6$, $p < 0.001$) and a significant interaction between conditions and tasks ($F(3,30) = 13.44$, $p < 0.001$). Effort increased, for all subjects, from the STA to the MOT condition. The contrast on this difference is significant for both the PT and RT case (PT: $t(30) = 8.10$, $p < 0.001$ and RT: $t(30) = 11.32$, $p < 0.001$). It can be concluded that the addition of motion significantly increased control effort. Note the difference in the values of the effort metric obtained for the PT and RT; the much lower values in the RT task is in agreement with the task instruction in the RT to use 'minimum effort'.

Cancellation in the CAN condition significantly decreased control effort (PT: $t(30) = 3.75$, $p < 0.001$ and RT: $t(30) = 3.44$, $p < 0.001$), for some subjects back to the no-motion level (STA). Only one subject (#2) showed higher effort in the RT-CAN condition with respect to the RT-MOT condition, for all other subjects effort decreased. Comparing the CAN and INC conditions, results for the PT showed a significant increase in effort ($t(30) = 2.76$, $p < 0.001$); in the RT, however, no significant difference was found ($t(30) = 0.59$, $p = 0.278$).

Although the control effort metrics show a larger spread than observed for the performance metric, overall the hypotheses are confirmed. Effort increases due to BDFT effects but can be significantly reduced when cancellation is active. For the PT, control effort was lower with congruent cancellation than with incongruent cancellation.

8.6.5   Performance-effort balance

In manual control tasks performance and effort are related, which makes performance and effort two coupled factors determining the control behavior. This is especially important when looking across conditions. If a subject keeps effort constant across conditions, changes in performance can be expected. It is also possible that a subject accepts a certain performance range and adjusts effort accordingly, etc.

To investigate the performance-effort balance the *difference* between

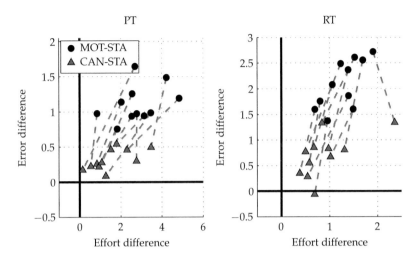

Figure 8.11: Difference in performance and effort, for each individual subject, between the MOT and STA conditions and between the CAN and STA conditions.

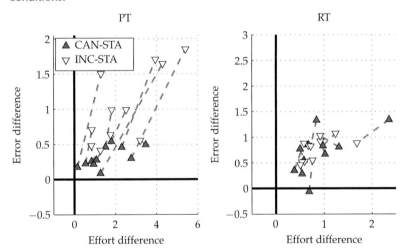

Figure 8.12: Difference in performance and effort, for each individual subject, between the CAN and STA conditions and between the INC and STA conditions.

metrics obtained in two conditions was calculated. Fig. 8.11 shows the difference between the STA and MOT condition (MOT-STA) and the STA and CAN condition (CAN-STA). For both cases, the STA condition was used as baseline, so the difference value is with respect to the value obtained in the STA condition. For the MOT-STA data all differences are positive, for both the error and effort metric. This implies that for each subject both the effort and error increased (i.e., performance decreased) when motion was added. Also all CAN-STA data is generally positive, but the increase in effort and error is much smaller. The lines between the data points connect the data belonging to the same subject. From these lines it becomes clear that for each subject the error decreased when cancellation was added (with respect to the MOT condition). For most subjects, also the effort decreased. The only exception is the subject plotted in the far upper right corner of the RT plot. This subject shows an increase in effort (subject #2). For the RT, the difference in effort is generally small, but a clear difference in performance is visible.

Fig. 8.12 allows to investigate the effect of congruent and incongruent cancellation. The PT plot shows that for most subjects, incongruent cancellation increased both effort and error, with respect to the CAN condition. This indicates that, for the PT, incongruent cancellation has a detrimental effect. As seen before, for the RT condition, this effect is not present. These results for this task indicate that incongruent cancellation, i.e., applying a PT model while the subject is performing an RT task, still results in similar performance and effort, both lower than without cancellation.

8.6.6 Straight HITS sections

In the straight HITS conditions, cancellation yielded a positive percentage for all conditions and all subjects. For PT-CAN, the cancellation level varied between 36.7% and 60.4% (average (μ) 49.3% and standard deviation (σ) of 9.0%). For RT-CAN the average is similar ($\mu = 50.9\%$) but the variability between subjects is considerably larger; P_{can} varies between 15.0% and 73.8% ($\sigma = 21.1\%$). The same subjects as in the CUR conditions (#1, #2, #3 and #9) showed the lowest cancellation results in the RT-CAN condition.

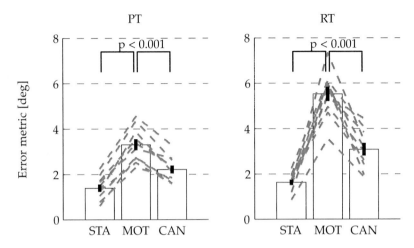

Figure 8.13: Tracking performance for the three conditions (PT and RT sections), straight HITS. The bars indicated the mean, the lines indicate the standard error of the mean. The dashed gray lines connect the mean values obtained for each subject.

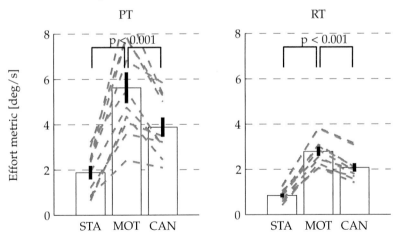

Figure 8.14: Control effort for the three conditions (PT and RT sections), straight HITS. The bars indicated the mean, the lines indicate the standard error of the mean. The dashed gray lines connect the mean values obtained for each subject.

Figures 8.13 and 8.14 show the tracking performance and control effort metrics, respectively. The trends in these metrics were the same at those found for the curved HITS conditions, with similar levels of statistical significance. For a detailed discussion the reader is referred to [Venrooij et al., 2014c].

In conclusion, results obtained from the STR tunnels lead to identical conclusions regarding the BDFT hypothesis and the cancellation hypothesis.

8.6.7 The closed-loop case

An interesting variation of this experiment would be to 'close the loop' and have the subjects actually control the accelerations of the simulator. One reason this was not done in this study is that in the closed-loop case it would not have been possible to obtain estimates of the B2FCL$^+$ and B2P dynamics, which are central in the analysis provided here. Another reason is that in the closed-loop case the accelerations would follow the vehicle movements and therefore not be nearly as strong as they were in this study. BDFT effects might still occur, but less frequently and less measurable. A realistic situation where more severe BDFT effects would occur in closed-loop vehicle control is when external turbulences act on the vehicle. A situation with considerable turbulence, i.e., where the turbulence-induced vehicle motions are much stronger than the commanded vehicle motions, would approximate the open-loop case studied here.

An important question is whether the cancellation would still work in closed-loop. It is reasonable to assume that it would, given that the accelerations are accurately known or measured and the following three conditions are met: the subject's neuromuscular settings are similar to those in the identification measurements; the identification measurement provided an accurate BDFT model; and the frequency spectrum of the closed-loop accelerations falls within the frequency spectrum used in the identification measurements. Note that these requirements also hold for an open-loop experiment. If these requirements are met then the BDFT model should closely approximate the involuntary BDFT effects induced by the platform's accelerations, whether these accelerations stem from an open-loop signal, turbulence, or closed-loop vehicle control.

8.7   Conclusions

From the results it can be concluded that the BDFT model described the measured BDFT dynamics – both the B2FCL$^+$ and B2P dynamics – well. Trends in all metrics correspond well with the expected effects of the motion disturbance and cancellation on tracking performance and control effort. Cancellation yielded positive cancellation percentages for all subjects, with a better and more robust performance in the PT as compared to the RT. The observation that for some subjects the RT-INC condition provided better cancellation results than the RT-CAN condition can be explained by the fact that these subjects expressed different behavior in the identification measurements and the cancellation experiment.

Regarding the hypotheses tested in the experiment, the following conclusions can be drawn:

- The BDFT hypothesis was confirmed: BDFT occurred in the motion condition (MOT) and significantly deteriorated tracking performance and increased control effort with respect to the static condition (STA).

- The cancellation hypothesis was confirmed: with cancellation (CAN) the tracking error and control effort were significantly lower than without the cancellation (MOT).

- The incongruency hypothesis was only partially confirmed: incongruent cancellation (INC) leads to lower performance and higher effort than obtained with congruent cancellation (CAN) for the PT only.

The latter point may raise the question what the added value is of having an admittance-adaptive approach at all, as the PT model seems to perform fine in both conditions. In fact, in an earlier publication Venrooij et al. [2011b], it was already predicted that "the 'safest' model choice appears to be the position task model, as it shows a partial reduction, even across tasks" (p. 1675). In the same publication it is added, however, that the level of cancellation that can be achieved with a PT model is lower than that which can be achieved with a task-specific model.

In the analysis it was shown that some subjects executed the RT task differently in the identification measurements and the cancellation experiment. This led to a poor match between the BDFT model and the actual BDFT dynamics for these subjects and this explains why the RT-INC showed better results than the RT-CAN condition for some subjects. This still illustrate the importance of using a BDFT model that matches the *actual* BDFT dynamics, i.e., the actual task that is performed.

In this study only two types of control tasks were studied, the PT and RT. Evidently, a human operator is capable of many different types of control behavior, each with its influence on the setting of the neuromuscular dynamics and thus on the BDFT dynamics. Studying the success of cancellation across a wider range of these possible settings is likely to show the benefit, if not the necessity, of an adaptive cancellation approach that matches the neuromuscular setting as closely as possible. This issue can only be properly addressed with *online* identification techniques that allow for real-time adaptation of the cancellation model to the current settings of the neuromuscular system. These methods are under development Mulder et al. [2011]; Tanaka et al. [2014].

Table 8.5: Model parameters for each subject.

$$H_{B2P}^{mod} = K H_B^1 H_B^2 H_B^3 H_B^4 H_B^5 H_{CD}$$

with $H_B^1 = (1 + 2\zeta_1/(2\pi f_1 s) + s^2/(2\pi f_1)^2)^{\gamma_1}$, etc.

and $\gamma_1 = -1, \quad \gamma_2 = +2, \quad \gamma_3 = -2, \quad \gamma_4 = +2$ and $\gamma_5 = -1$

PT MODEL

subj.	K [-]	f_1 [Hz]	f_2 [Hz]	f_3 [Hz]	f_4 [Hz]	f_5 [Hz]	ζ_1 [-]	ζ_2 [-]	ζ_3 [-]	ζ_4 [-]	ζ_5 [-]
# 1	1.431	0.976	1.247	2.772	6.443	7.037	0.666	0.407	0.366	0.093	0.051
# 2	1.059	0.800	1.177	3.321	5.803	5.602	0.535	0.399	0.345	0.034	0.010
# 3	0.699	0.835	1.017	2.681	6.090	7.004	0.217	0.346	0.374	0.214	0.143
# 4	0.784	1.067	1.131	2.977	5.834	5.404	0.133	0.166	0.450	0.170	0.079
# 5	0.994	1.288	1.256	2.996	9.101	10.571	0.485	0.266	0.525	0.193	0.113
# 6	1.058	1.119	1.162	3.171	6.449	5.007	0.153	0.175	0.548	0.211	0.201
# 7	0.510	1.500	1.252	3.323	6.961	5.586	0.614	0.309	0.480	0.360	0.187
# 8	1.098	1.172	1.278	3.578	7.210	5.000	0.728	0.375	0.522	0.609	1.000
# 9	1.939	0.585	1.229	3.426	5.659	5.469	0.763	0.529	0.370	0.048	0.020
# 10	1.285	0.756	1.147	3.106	6.496	6.965	0.756	0.439	0.295	0.045	0.018
# 11	0.610	1.287	1.192	3.893	9.660	10.170	0.565	0.290	0.524	0.034	0.010

RT MODEL

subj.	K [-]	f_1 [Hz]	f_2 [Hz]	f_3 [Hz]	f_4 [Hz]	f_5 [Hz]	ζ_1 [-]	ζ_2 [-]	ζ_3 [-]	ζ_4 [-]	ζ_5 [-]
# 1	1.823	0.925	1.225	2.032	4.798	5.686	0.252	0.353	0.539	0.304	0.200
# 2	1.426	0.872	1.367	2.270	4.266	5.000	0.207	0.355	0.450	0.408	0.267
# 3	1.039	0.838	1.114	2.462	6.361	7.618	0.197	0.456	0.649	0.294	0.170
# 4	0.749	0.916	1.065	2.814	6.618	6.991	0.244	0.298	0.478	0.119	0.033
# 5	1.037	0.916	1.145	2.581	6.695	7.839	0.322	0.365	0.502	0.223	0.117
# 6	1.221	0.481	1.159	3.275	5.811	6.566	0.736	0.992	0.410	0.093	0.070
# 7	0.820	1.008	1.073	2.551	6.762	7.592	0.115	0.177	0.533	0.217	0.145
# 8	1.288	0.942	1.126	2.409	5.291	5.000	0.214	0.265	0.535	0.281	0.328
# 9	2.540	0.979	1.147	2.056	6.740	7.738	0.208	0.287	0.919	0.362	0.188
# 10	1.070	1.064	1.151	2.412	6.007	6.485	0.169	0.202	0.427	0.049	0.014
# 11	0.775	1.018	1.095	2.617	6.754	7.239	0.169	0.215	0.551	0.152	0.061

CHAPTER 9

Discussion

9.1 General discussion of the results

The previous chapters have focused on several aspects of biodynamic feedthrough, particularly measuring, analyzing, modeling and mitigating its effects. In this section we will expand our focus and discuss the progress that was made in these four domains in a larger perspective.

9.1.1 Measuring biodynamic feedthrough

Limitations of the measurement method

The measurement method that allows for a simultaneous measurement of neuromuscular admittance and biodynamic feedthrough, presented in Chapter 2, was used extensively in the experiments conducted in the context of this thesis. The fact that experiments with different subjects, conditions and setups have shown to produce reliable and repeatable results is an indication of the quality and usefulness of the method. However, there are some weaknesses that should be kept in mind as well.

First of all, the system identification techniques that were used are inherently limited to linear time-invariant (LTI) systems. With carefully designed experimental conditions one can justify the assumption that the measured dynamics are linear around a constant operating point and are time-invariant for the duration of the measurement. A similar assumption of linearity was made in several previous studies ([Donati and Bonthoux, 1983; Lewis and Griffin, 1979]). The validity of this assumption can be checked using the squared coherence of the obtained data. Nonetheless, this assumption is a strongly limiting one and sets strict boundaries on the allowable deviations from the experimental conditions used in this thesis.

In the experiments, the dynamics of the neuromuscular system was structurally varied using the following three control tasks (also referred to as the classical tasks): position task (PT), force task (FT) and relax task (RT) [Abbink, 2006]. The classical tasks have been used in numerous studies to investigate the variability of the neuromuscular system, e.g., [Abbink, 2006; Lasschuit et al., 2008; Mugge et al., 2009]. The control tasks served first and foremost to elicit a certain setting of the neuromuscular system, ranging from 'stiff' to 'compliant', but also to keep this setting as constant as possible for the duration of the measurement. If one wants to measure biodynamic feedthrough (and/or admittance) for a situation in which the neuromuscular setting varies over time, as is the case in many natural control tasks, not only different control task instructions but also different system identification techniques are required. Furthering the investigation into these techniques is highly relevant as they allow for studying more realistic and possibly more relevant control behavior. Also, more advanced methods may provide a valuable online estimate instead of just a post-hoc result (see, e.g., [Katzourakis et al., 2013; Mulder et al., 2011]).

Measuring the relax task

On several occasions there were difficulties in measuring the admittance in the relax task (RT). It was observed that the admittance measured for this task showed a gain dependency, i.e., with increasing force disturbance gain the admittance magnitude increased (Chapter 4). Furthermore, the results obtained in conditions where the motion disturbance was active showed an increase in RT admittance magnitude with respect to the static case (Chapter 3).

For both observations a possible and plausible cause is the influence of gravity on the admittance estimate. For the stick-type control devices used in this study, gravity has the effect of magnifying deviations from the center position of the control device by pulling the arm down. The larger the deflections from the center position, the larger the role of the downward pointing gravitational force component. By increasing the force disturbance gain or by adding a motion disturbance signal, the resulting increase in control device deflections was likely further magnified by this gravity effect.

As the neuromuscular admittance is the dynamical relationship between force input and position output, the magnification of the position outputs resulted in an increase in the estimated admittance magnitude. Hence, the arm appears to be more compliant than it actually is. The fact that this effect only occurs in the relax task and not in the other two tasks can be explained by the passive behavior elicited in the RT. In this passive state the magnifying effect of gravity is not counteracted by active control inputs. The issue of measuring admittance and BDFT in the relax task, and especially the 'gravity hypothesis' suggested here, deserves further investigation.

How representative are the classical tasks?

An important question is how representative the classical tasks are for manual control in actual vehicles. The control tasks were designed to elicit maximally stiff, maximally compliant and maximally passive behavior. One may wonder how often these maxima are actually reached in real-life situations.

As was shown in Chapter 8, not all subjects were able to combine the RT setting with a vehicle control task. From the data it was concluded they exhibited slightly stiffer behavior than when performing the RT alone. This is not a surprising finding (the muscle activity required for voluntary control increases the visco-elasticity of the limb which reduces the limb's admittance), but it does raise the question whether the RT is a task that can be performed in combination with vehicle control at all.

Similar questions can be raised for the other classical tasks. It is rather unlikely that a well-trained helicopter pilot would adjust his neuromuscular dynamics to the maximally stiff settings as elicited in the position task (PT). It is hard to imagine a conditions in which an excavator operator sets his neuromuscular dynamics to be as complaint as when performing a force task (FT). In general, the control tasks as used in this thesis are unlikely to naturally occur during actual vehicle control. Still there are several valid arguments in favor of using them.

First of all, they have shown across numerous studies to produce repeatable results for which the LTI assumption mentioned earlier is valid. Furthermore, rather than viewing the control tasks as representative settings, one could (or should) interpret them as the 'extreme cases', exposing the boundaries of possible variation. They allow for studying the amount of variability humans are able to generate and how this variability influences other aspects of vehicle control, such as biodynamic feedthrough.

Using neuromuscular adaptation as BDFT mitigation technique

If the required identification techniques become available it would be worthwhile to investigate admittance and BDFT dynamics that occur 'between' the extreme settings of the classical tasks investigated here. The results of such a study would deepen our knowledge of admittance and BDFT dynamics and provide an interesting source of data for the further development and validation of admittance and BDFT models. Besides allowing us to study more representative settings for actual vehicle control tasks than the classical tasks, they may also provide an additional source of BDFT mitigation.

The results presented in Chapter 2 showed that there is no setting for which BDFT is minimal for all frequencies. However, it is possible that a particular setting does provide some sort of minimal feedthrough and thus minimizes BDFT problems. The particulars of this setting, which will differ depending on vehicle dynamics, control device dynamics, external disturbances, etc., may be used in training human operators to mitigate BDFT themselves, i.e., without additional equipment. Helicopter pilots are typically trained to apply a loose grip and relaxed handling of the control devices. Amongst other benefits, this helps them to minimize the feedthrough of vehicle accelerations. Similar instructions may help operators of other vehicles to attain the optimal neuromuscular setting to minimize BDFT. In some cases it may be beneficial to increase grip (reducing BDFT at lower frequencies), in others cases to release (reducing BDFT at higher frequencies). Possibly, the optimal setting is an unusual combination of grip and co-contraction settings (such as those studied in [Nakamura et al., 2011]).

Biodynamic feedthrough in closed-loop

The work in this thesis has been limited to open-loop BDFT systems. Many practical BDFT problems, however, are closed-loop in nature. It is reasonable to assume that many of the results obtained in open-loop, e.g., the influence of an armrest or control device dynamics, are largely transferable to closed-loop situations. The supportive reasoning is based on the observation that the feedthrough of accelerations is involuntary and occurs independently of whether these accelerations are under control of the operator or not. The BDFT models that were developed in this thesis remain largely applicable in the closed-loop case, by closing the OL-CL switch and adding vehicle dynamics to the model. Also the mitigation techniques, that were studied here in open-loop, are expected to be largely transferable to closed-loop systems.

Still, there are several important aspects of BDFT that are unique to closed-loop systems, such as the occurrence of oscillations, which were not studied in this thesis. Closed-loop BDFT systems pose some additional challenges in terms of the system identification techniques required to study them. The identification techniques used in this thesis are – due to the strict requirements on the frequency contents of the acceleration disturbance signal – not suitable for application in closed-loop situations. The study of specific closed-loop BDFT problems, such as BDFT instabilities, is an important future research direction.

9.1.2 Analyzing biodynamic feedthrough

The framework for BDFT analysis, proposed in Chapter 3, has the potential of becoming a valuable contribution to the fragmented research field of BDFT. The introduction of a framework is an important step towards creating some common ground between different approaches and facilitate communication between researchers and comparison between studies.

Earlier attempts at proposing a framework, such as the one in [Lewis and Griffin, 1976], have often had only a limited lasting impact on the way future research was conducted. A hurdle that every unifying effort needs to overcome is the inertia of the established research doctrines: a change in nomenclature and definitions is never adopted lightly, as it requires a change in mindset and a shift away from what one is comfortable with. The potential benefit of the framework can only come to full fruition if it is used, evaluated, criticized, discussed and improved by a sufficiently large body of users.

One of the strengths of the framework is its theoretical foundation, the result of several years of research, which was thoroughly validated and which represents the state-of-the-art in knowledge on BDFT. What the framework is currently still lacking is 'experience in the field'. The application of the framework to diverse BDFT problems is essential in unveiling the framework's strengths and weaknesses. The main challenge is now to spread the framework amongst researchers with various backgrounds and research interests. Hopefully, this will not only lead to an increase in congruency across studies but also to numerous adaptations to and extensions of the framework, increasing its usability and acceptance.

9.1.3   Modeling biodynamic feedthrough

A bias towards practical models

Models capturing BDFT effects have been around for decades, but none of them seems to be particularly well established or accepted. Compared to, e.g., the domain of manual control theory, where McRuer's Crossover Model (COM) has been one of the central models for decades, the BDFT domain has shown much less unity between the various modeling efforts. The widespread use of the COM (or one of its many adaptations) illustrates that an established model can facilitate progress as not every researcher in the field has to design his/her own models.

The BDFT models that do exist in the literature seem to fall short in either fidelity or usability [Griffin, 2001]. For a good model both fidelity and usability are important traits, but practice has shown a bias towards preferring the latter. To illustrate this, let us consider a physical model proposed by Jex and Magdaleno [1978], which utilizes "a homologous or life-like representation of major body segments in their orientations" (p. 306) to construct a model containing over 70 parameters. Let's compare that model to the black box model proposed by Mayo [1989], which utilizes "a second order analytical transfer function fit" (p. 5) to construct a model containing only 4 parameters. The two models illustrate two extremes. The physical model excels in physiological fidelity: it incorporates head bobbing on an articulated neck, sliding hip and rocking chest, amongst many other features [Jex and Magdaleno, 1978]. Such detail is necessary to increase our *theoretical understanding* of biodynamic feedthrough on a fundamental level. The black box model, on the other hand, excels in practical usability: the two parameter sets that were obtained for two different body types in [Mayo, 1989] make this model directly applicable. Such a model does not necessarily aid in understanding BDFT any further, but is much more suited for a *practical implementation*. Regarding the drawbacks of the two approaches: the physical model is often hard to implement and use, black box models lack a physical interpretation and are often more limited to specific applications [Griffin, 2001; Venrooij et al., 2013b].

Without going into further model details or possible objections regarding the way they were obtained, let us look at how the models were used by the scientific community: where the physical model of [Jex and Magdaleno, 1978] has no known implementation in other studies, Mayo's black box model has been used frequently in a number of different studies, also recently (e.g. in [Dieterich et al., 2008; Gennaretti et al., 2013; Masarati et al., 2007; Mattaboni et al., 2008; Quaranta et al., 2013; Serafini et al., 2008]). In fact, Mayo's model seems to be the only model that is regularly used in the BDFT (rotorcraft) community, despite its known limitations.

This comparison between the two models serves merely as an illustrative example, but the lesson one can draw from this is that the direct practical applicability of a BDFT model are important traits. In this case, and probably several others, the research community has shown to have a bias towards such models. Future BDFT modeling efforts should therefore be directed to obtain models of limited complexity, without compromising fidelity.

This does not mean that complex and elaborate physical models should be avoided all together. On the contrary. Such models can provide a level of detail and increased insight that cannot be matched by other model types. The complexity of implementing and using such models, however, is likely to reduce the effective user group to a small number of specialists. In order to ensure a BDFT model is beneficial to a larger audience, including, e.g., rotorcraft designers with limited understanding of BDFT, a simpler and more practical model is likely to be more successful.

Strengths and weaknesses of the physical model

The implementation and interpretation of the physical model, proposed in Chapter 4, may be a daunting task for many researchers. As argued above, a researcher in the rotorcraft domain, interested in BDFT effects in a particular helicopter, is unlikely to have a strong background in neuromuscular admittance modeling. In this case the additional insight the physical model provides probably does not weigh up against the complexity of implementing, estimating its parameters and interpreting its results.

In contrast, for those familiar with neuromuscular identification and modeling the implementation of this model will pose much less of a problem. In fact, the physical BDFT model is a direct extension of a well-known and well-documented neuromuscular admittance model [Abbink, 2006; Mugge et al., 2009]. The effort needed to obtain a BDFT model from the admittance model is limited to the addition of only several parameters.

The strength of the model lies in what it can reveal regarding the interaction between admittance and BDFT. The development of the model that was presented in this thesis laid the ground work for a future study into the details of this highly relevant relationship. For example, a question that was only briefly touched upon in this thesis is whether BDFT dynamics can be *predicted* from admittance by exploiting relationships between model parameters. There are indications this is possible from the results reported in Chapter 4. If this indeed would prove to be the case this would be quite a leap forward in BDFT research, as methods to measure admittance have matured considerably in recent years. Establishing a strong relationship between the two dynamics would allow BDFT research to benefit from, for example, the progress that has recently been made in online admittance estimation techniques (e.g. [Katzourakis et al., 2013; Mulder et al., 2011]).

Strengths and weaknesses of the mathematical model

The mathematical BDFT model, provided in Chapter 5, offers a much more practical and at least equally accurate alternative to the physical BDFT model. The model is likely to be the most versatile BDFT model that is currently available: it accurately describes BDFT dynamics for three different directions and three different settings of the neuromuscular system. No other known BDFT model matches that scope. The model was quite thoroughly put to the test when using it in the mitigation experiment described in Chapter 8. The fact that for each subject accurate models could be obtained within minutes is a sign of the robustness of the model and the parameter estimation process.

The model has so far only been used for studies using typical heli-copter control devices. It would be rewarding to put the model to use in other setups with different control devices. It is likely that this requires adaptations to, e.g., the number and type of base func-tions used.

A weakness of the mathematical model is its dependency on con-trol device dynamics. By making use of B2FCL dynamics instead of B2P dynamics, the dependency was changed from a strong one to a weak one, but it does not remove its influence completely. The two model-types that do not have such a dependency are the phys-ical BDFT model (in which the control device properties are model parameters) and a B2FOL model (in which there is no dependency on control device dynamics).

A major disadvantage of a B2FOL model is that the conversion to B2FCL or B2P dynamics requires knowledge (i.e., a model) of the neuromuscular admittance. This makes using a B2FOL model con-siderably more difficult, negating the benefit of control device dy-namics independence.

Evaluation of the asymptote modeling approach

The proposed method of obtaining the model structure through asymptote modeling is novel and is likely to have merit for other modeling problems as well. The method was thus far only applied to the modeling of B2FCL dynamics, but there is no reason to as-sume that its benefits are limited to this type of dynamics. For example, it would be interesting to attempt constructing an admit-tance model using the asymptote modeling approach.

Just as with any other modeling technique, one needs to gain some understanding in the modeling problem and the available tools be-fore a satisfactory result can be obtained. The selection of the right (combination of) base functions may take some experience before the method can be successfully applied, especially in more com-plex modeling problems.

In comparison with other modeling techniques, such as the phys-ical modeling approach or a criterion-based selection of transfer function orders, the asymptote modeling approach offers several benefits. First of all, the approach allows a researcher to accurately focus the model on features that he or she thinks are of importance. Also, this focus can easily be refined in a later stage by adding (or removing) one or more base functions. The model's robustness to noise can easily be improved by adding one or several base functions that attenuate, e.g., high frequency noise.

The fact that the model parameters retain a mathematical interpretation has two major advantages: the first is that the model has a high degree of transparency, as the role of each parameter in the model remains clear. The second advantage is that this makes the parameter estimation procedure relatively easy and straightforward, for both manual and automatic estimators. In the case demonstrated in this thesis – a B2FCL model with 16 parameters – the parameter estimation could have been largely done by hand, illustrating the transparency of the model. Better results are often obtained using a automated iterative optimization techniques, and also these methods benefit from the unambiguous interpretation of the parameters. The way asymptote modeling was introduced in this thesis still involved some 'manual labor', such as the selection of the type and number of base functions. There is good reason to assume that most of these steps can be automated, in order to save time and/or improve performance. One elegant aspect of asymptote modeling is, however, that it allows a researcher to have a high degree of influence on the modeling process. There are many cases in which even the most advanced modeling algorithms cannot compete against a researcher's extensive experience with the dynamics that need to be modeled, and, maybe more importantly, the researcher's knowledge regarding the future use of the model. Asymptote modeling allows for successfully converting such knowledge into a model that, e.g., strikes the right balance between complexity and accuracy.

The asymptote modeling approach has only been applied here to the problem of B2FCL modeling, in which it proved very successful. Applying the same technique in other modeling challenges is essential to further develop the approach.

9.1.4 Mitigating biodynamic feedthrough

The armrest

Amongst the many attempts at BDFT mitigation, the armrest seems often to have been ignored (exception are studies by Torle [1965] and McLeod and Griffin [1995]). This is surprising, as it offers a straightforward solution to this rather complex problem. A possible cause for this may have been the lack of measurement techniques with which its influence can be shown in detail. The measurement techniques developed in this thesis allow for measuring the effectiveness of the armrest on BDFT dynamics in more detail than was possible before.

The results show that the addition of an armrest reduces the occurrence of BDFT considerably (for frequencies below 2 Hz to about 50% to 30% of the magnitude without armrest). It is likely that in many practical cases an armrest may be all that is required to reduce BDFT sufficiently to obtain an adequate task performance. This result makes the armrest a priority candidate to be considered when mitigating BDFT in practice.

Strengths and weaknesses of model-based BDFT mitigation approach

The results of the admittance-adaptive model-based BDFT mitigation experiment, presented in Chapter 8, leave little doubt about its effectiveness: the approach was successful in largely removing BDFT for the conditions that were investigated.

Model-based BDFT cancellation is likely to be the most versatile and accurate BDFT mitigation approach available. An important motivation to use the model-based signal cancellation approach in this thesis – and not any other approach – was that it allowed to put a significant part of the new knowledge and insights to the test. Furthermore, the method is preferred over many other approaches, as it allows for BDFT mitigation without compromising any other aspect of the human-machine system, i.e., it does not require adaptations to vehicle dynamics, control device layout or human control behavior. At the current moment, model-based BDFT mitigation is the only method for which it has been demonstrated that it can cope with both between-subject and within-subject BDFT variability.

The implementation of a model-based BDFT mitigation approach is not an easy task, as was illustrated in Chapter 6 and Chapter 8. For example, not matching the neuromuscular setting of the operator can to lead to a reduced quality of the mitigation, possibly leading to a situation where the control performance is further deteriorated instead of improved.

A weakness of the way the model-based BDFT mitigation approach was implemented in this thesis is the manner in which the identification measurements were executed: in order to minimize the influence of visual-based cognitive corrective control inputs, the visual information (the HITS) was removed in the identification measurements. This led to differences in neuromuscular dynamics between the identification measurements and the cancellation experiment, which in turn led to differences between the modeled and actual BDFT dynamics, reducing the quality of the cancellation.

One way to deal with this would be to perform a thorough investigation into the effects of cognitive corrective control actions. This would allow for conducting the identification measurements in the same conditions as the cancellation experiment (i.e., with the HITS visually present) and correct these measurement for the influence of visual-based corrective control actions afterwards. Next to its usefulness in improving the quality of BDFT mitigation models, this knowledge would also increase our understanding on how humans are able to use visual information to correct for BDFT effect, which is a highly relevant topic in its own right.

The feasibility of model-based BDFT mitigation

An important question is whether the model-based cancellation approach is currently feasible in practice. At the moment, we must conclude that, despite its potential benefits, a practical implementation of the method is still out of reach for several reasons, of which only the two most important are highlighted here.

First of all, the method has not matured enough to make a practical implementation possible. Model-based cancellation is one of the most complex methods of mitigating BDFT that we have at our disposal. The implementation of the method in this thesis and other works has left many questions unanswered. More research is required to refine the mitigation approach.

Secondly, the results presented in this thesis have shown that the variability of the human neuromuscular system has an important influence, calling for an adaptive cancellation approach. Despite the progress that was made in recent years, the current state-of-the-art does not yet provide the required methods to obtain a reliable online estimate of either admittance or BDFT. Future research may provide ways to derive admittance or BDFT estimates in an online fashion or from indirect metrics, such as grip or electromyography (EMG) data. At the current moment, however, the reliable and fast online identification of BDFT dynamics remains an unsatisfied requirement for the model-based cancellation approach to be successful in practice.

A definitive BDFT mitigation approach

Even if the technical difficulties with the model-based BDFT mitigation approach are largely overcome (which is to be expected), it is not at all likely that this approach will become the definitive approach to BDFT mitigation that will be widely accepted as 'the best way' of canceling BDFT. Instead, simpler strategies, such as the armrest, are likely to remain viable alternatives to the more complex approaches, even if the latter can be shown to be in superior in their effectiveness.

The reason for this is what can be called the 'efficiency' of an approach. A more efficient approach is the one that reduces BDFT sufficiently to allow for an acceptable task performance at a lower cost. Costs refer not only to financial costs, but also to complexity of the approach, the time its implementation requires, the sacrifices it involves in other aspects of vehicle control, etc.

In many respects, an armrest can be considered to be an efficient approach, and model-based BDFT cancellation a less efficient approach. For the different occurrences of BDFT across vehicles it will be the balance between required effectiveness and efficiency that will determine the optimal mitigation approach. Appendix B provides some guidelines on how the selection of the optimal approach for a practical BDFT problem can be performed.

9.2   Relationship to previous works

In order to properly weigh the contributions of the work presented in this thesis, it is important to establish relationships with what has been done previously. This section aims at establishing such relationships to aid in identifying the (dis)agreement between the findings presented here and in other studies.

Different ways of defining biodynamic feedthrough and measuring its effects

Many different identifiers have been used for BDFT (related) phenomena. One example is "vibration feedthrough", which was a common term in many earlier BDFT related studies (e.g., [Allen et al., 1973; Jex and Magdaleno, 1978; Lewis and Griffin, 1976, 1979; McLeod and Griffin, 1989]). In those studies, vibration feedthrough was usually defined as the feedthrough of motion disturbances to controlled element (CE) output. It is important to note that, hence, vibration feedthrough includes the CE dynamics, while biodynamic feedthrough does not. This difference is, as we will see, important when comparing results between studies.

There is also considerable variability between studies in how BDFT effects are measured and reported. In many previous studies the analysis focused on identifying "the difference sources of vibration-related tracking errors" and their relative importance. The (root) mean square error was often used as performance metric, usually partitioned into input-correlated error, vibration-correlated error and remnant (see, e.g., [Allen et al., 1973; Lewis and Griffin, 1979]). Such an analysis results in a single value for the total error and single values for each contribution.

In this thesis, BDFT effects were usually reported through frequency response functions (FRFs), such as shown in Fig. 2.8. Clearly, the

FRF result obtained through this approach is not directly comparable to the single values that many other studies report. The FRF shows the amount of involuntary control *input* per unit of acceleration, for a *range of frequencies*, while many studies show the effect of vibration on control *output* (in the form of tracking error) through a *single value*. The FRF is particularly helpful in understanding the biomechanical aspect of how motion disturbances interfere with manual control. What it does not (directly) show, however, is how BDFT impacts the control of a system, say, a helicopter.

To illustrate why it is important to be aware of different ways of defining biodynamic feedthrough and measuring its effects, let us look at a study by Allen et al. [1973]. In this study, the vibration-correlated error (i.e., the vibration feedthrough) was calculated and it was observed that it was not a big contributor to the total control error. The addition of a vibration disturbance was found to mainly increase the remnant (see Fig. 25b in [Allen et al., 1973]), which suggests that vibration feedthrough only played a minor role in the tracking performance. Similar conclusions were drawn in [Lewis and Griffin, 1976] and [McLeod and Griffin, 1988]. In addition to calculating the contribution of the vibration to the error, the contribution of the vibration to the control signal (i.e., the BDFT) was also determined in [Allen et al., 1973]. Interestingly enough, it was observed that the vibration-correlated contribution to the control device input dominated the control signal (see Fig. 26b) and was many times larger than the remnant. Using the results presented in this thesis we can further increase our understanding of these seemingly contradictory observations regarding the contribution of vibration to error and control input.

It was remarked in [Allen et al., 1973] and [McLeod and Griffin, 1989] that vibration feedthrough is attenuated at higher frequencies for higher-order CE dynamics ("every integration in the system attenuates breakthrough [i.e., vibration feedthrough] at the system output by 6 dB for each doubling of the frequency" [McLeod and Griffin, 1988]). This suggests that the BDFT effect on the eventual state of the CE may be small, possibly unimportant, but it should be noted that the validity of such a conclusion hinges on the CE dynamics that were used. The CE dynamics in practical situations, e.g. in a helicopter, can contain complex features such as resonance peaks and structural modes, which may make BDFT a factor that should be reckoned with. By confining an analysis to vibration feedthrough with simplified CE dynamics, the effects of BDFT may be obscured and lead to a dismissal of BDFT as an insignificant contribution, which is not always justified. The occurrences of BDFT in actual vehicles (such as reported in [Walden, 2007]) or the beneficial effects of canceling BDFT that were reported in [Sövényi and Gillespie, 2007] and Chapter 8 of this thesis show that studying BDFT is indeed important.

The focus of this thesis was on BDFT and hence the work does not contribute significantly to understanding the other mechanisms through which motion disturbances can interfere with manual control (such as the effects of visual blurring or the increase of motor remnant). The methods that were developed in Chapter 2 can be used to contribute in those directions as well. The frequency decomposition technique (Section 2.4.7) can be used to quantify the relative magnitude and relevance of different mechanisms in very similar ways as was done in earlier biomechanical studies. An important benefit that this technique offers over what is used in other studies is the insight in the frequency spectrum that the results provide (see Fig. 2.9).

The influence of the control device dynamics

Several previous studies have investigated the influence of control device dynamics on BDFT (related) phenomena. Examples are [Jewell and Citurs, 1984; Lewis and Griffin, 1976, 1977, 1979; Torle, 1965] and more recently [Zaichick et al., 2012]. Next to variations in location, control gain and device type, also the influence of the control device dynamics were varied. Using the results presented in this thesis, some of these results can be better understood.

Several studies have investigated the influence of isometric sticks (or stiff sticks, which do not move and have applied force as output), isotonic sticks (which do not offer resistance to movement and have position as output) and spring(-centered) sticks (which resist movement proportionally to displacement and often include some damping). In the current thesis, research was done using isometric (stiff) sticks and spring sticks.

The results in Section 3.7.2 showed how the influence of isometric sticks and spring sticks on BDFT can be unified. It was shown how the B2P dynamics, measured using a spring stick, could be converted to B2FOL dynamics. The fact that the results were almost identical to the B2FOL dynamics that were directly measured using an isometric stick showed that a relationship between the effects of motion disturbances on isometric and spring sticks has been established. Furthermore, in Section 3.9 it was shown how the BDFT effects measured with one setting of spring stick dynamics can be converted to the those for other spring stick dynamics. This further illustrates that the influence of the control device dynamics on BDFT effects is now thoroughly understood.

In several studies it was reported that, without motion disturbances present, isometric sticks provided superior performance, but they seem more sensitive than isotonic or spring sticks to BDFT effects, especially at higher frequencies (> 5 Hz) [Lewis and Griffin, 1977, 1978c]. Similarly, Allen et al. [1973] reported that a spring stick resulted in "considerable high-frequency attenuation", while an isometric stick resulted in "relatively wide band feedthrough". Looking at the B2FOL dynamics (representing the involuntary force applied to an isometric stick) and the B2P dynamics (representing the involuntary deflection of a spring stick) these findings can be further explained.

Comparing B2P dynamics with B2FOL dynamics (e.g., as shown in Fig. 3.11 or Fig. 3.12) two important features emerge: (i) the magnitude of the B2FOL dynamics has a different unit and is one to several orders of magnitude larger than the magnitude of the B2P dynamics and (ii) the attenuation at higher frequencies is much greater for the B2P dynamics than the B2FOL dynamics, especially above 5 Hz. The first feature highlights that when comparing an isometric stick with a spring stick, it is important to realize that they have different outputs. It is therefore that any comparisons between such devices should be done with caution due to the difficulties in equating the gains of these controls [McLeod and Griffin, 1989]. The second feature explains the difference that was found regarding the sensitivity of isometric sticks at higher frequencies: while the B2P dynamics are strongly attenuated (due to the control device dynamics) the B2FOL dynamics are much less dependent on frequency.

The fact that isometric sticks were found to be more vulnerable to BDFT effects may have prevented their implementation in several vehicles, even though it was often shown that isometric sticks provide superior performance in static conditions. Therefore, it is worthwhile to investigate the possibilities of mitigating BDFT for isometric sticks. Using an approach very similar to the one outlined in Chapter 8, one could construct an admittance-adaptive model-based BDFT cancellation system. Through asymptote modeling, one can create a B2FOL model which predicts the involuntary BDFT-induced forces applied to the isometric control device. A successful reduction of BDFT effects for an isometric control device may show that isometric sticks retain their superior tracking performance even in motion conditions, making them preferable over other types of control devices, as long as the BDFT effects are effectively canceled.

The influence of the neuromuscular admittance

Amongst all the identified influencing factors, neuromuscular dynamics have often been mentioned as a possible source of variation, but has never been systematically studied. This is illustrated by the literature reviews provided in [Lewis and Griffin, 1978c] and [McLeod and Griffin, 1989], which discuss works that investigated a broad range of influencing factors, including workload and fatigue, but not the (possible) influence of neuromuscular dynamics or adaptation. The results in this thesis show that neuromuscular adaptation does indeed play a major role in the occurrence of BDFT effects which should not be overlooked. Even a complete understanding of the influence of seating dynamics or control device dynamics on BDFT does nothing to reduce the variability that is introduced by the body dynamics of different human operators or the same human operator under different circumstances.

The importance of accounting for variability within and between subjects has been recognized in many studies, such as [Griffin, 1978, 1981, 2001; Jex, 1972]. Most studies, however, mainly considered the so-called inter-subject variability, i.e., the variability that exists between subjects. For example, in [Mayo, 1989] two parameter sets were proposed, one for the ectomorphic somatotype (slim bone structure and muscle build) and one for the mesomorphic somatotype (athletic bone structure and muscle build). Such a distinction is reasonable as significant correlations were found between the vibration response and body size [Griffin, 1978]. Between-subject variability was also encountered in [Sövényi and Gillespie, 2007] and handled by using different models 'tailored' to each subject.

Although the *importance* of accounting for variability in BDFT has been widely recognized, the *methods* to actually study and understand this variability seem to have been largely lacking. One of the contributions of the work presented in this thesis is that it offers such methods.

What the results in this thesis fail to show, however, is a correlation between the performance of a BDFT model and the somatotype (physical body type) of the subjects (see Chapter 5). One would expect that an averaged model would perform well for a subject with a somatotype close to the 'average' somatotype that the model is based on and performance would decrease if the somatotype is very different. No such systematic relationships were found in the data. In [Venrooij et al., 2013b] it was observed that the BDFT data for the somatotypic groups showed only minor differences with respect to each other and the grand average BDFT. This does not agree with earlier findings and warrants further investigation.

Modeling biodynamic feedthrough

There are many biodynamic models available in existing literature, of which some deal with biodynamic feedthrough effects (e.g., [Allen et al., 1973; Hess, 2010; Jex and Magdaleno, 1978; Lewis and Griffin, 1976; Mayo, 1989; Sirouspour and Salcudean, 2003; Sövényi, 2005]). Despite this seemingly abundant spectrum of models, only few are re-used or refined in follow-up studies. Many models, it seems, are left abandoned after they are proposed. There is a myriad of reasons for this, of which the three most prominent seem to be that (i) the complexity of biodynamic problems often results in very complex unwieldy models, (ii) the applicability of both simple and complex models is usually limited to specific situations, and (iii) many models that are proposed in the literature are poorly documented and/or validated [Griffin, 2001]. Practice has shown that the complexity of a model can be a hindrance in its implementation. As was already argued in Section 9.1.3, there seems to be a bias to simple, practical models, although such models may actually not be the most accurate models available.

Griffin [2001] provides fair critique on how many biodynamic models that are available in the literature were conceived and validated. It is stated that the most useful model for any application is likely to be the simplest model that provides a sufficiently accurate prediction of the response of interest and that for some well-publicized models a sensitivity analysis would have shown that they are unnecessarily complex. Quality checklists are proposed as guidelines on how to construct and validate such models.

When applying the checklists proposed in [Griffin, 2001] on the two models proposed in this thesis, i.e., the mechanistic model in Chapter 4 and the qualitative model in Chapter 5, it can be concluded that models meet the majority of the requirements, but not all. This likely holds for many, if not all, BDFT models that are currently available in the literature. The overall quality of BDFT models would improve if they would more strictly adhere to a unified set of requirements, such as those proposed in [Griffin, 2001]. Clearly stating the assumptions, limitations and scope of application of a model, and making the data that was used to develop and validate the models readily available, would allow other researchers to implement the model, test it independently and adjust it for their purposes. This would increase the (re)usability, scope and quality of BDFT models.

Mitigating biodynamic feedthrough

Several studies were devoted to BDFT mitigation, such as [Dimasi et al., 1972; Humphreys et al., 2014; Sirouspour and Salcudean, 2003; Sövényi and Gillespie, 2007; Velger et al., 1988]. Comparing the results of those studies with the results presented in this thesis is difficult, due to differences in the applications the mitigation methods were developed for, but mainly due to differences in how the mitigation effects were quantified. Already between chapters in this thesis different metrics were used to evaluate the effect of the armrest and the quality of the model-based signal cancellation approach. Guidelines on how to best quantify the quality of a mitigation approach would, when accepted and adhered to, aid in weighing the benefits and disadvantages of different approaches. Such guidelines could be partially derived from the above discussed modeling guidelines. When quantifying the effect of any mitigation approach it is unlikely that a single value will be sufficiently informative to allow for a fair comparison. Indexing the various ways in which the success of BDFT mitigation approaches has been quantified in existing literature would provide an interesting and fairly comprehensive overview of the possible methods that we have at our disposal. New insights, such as the ones provided in this thesis, should also be 'translated' into metrics or requirements. For example, the results in this thesis show that the effect of neuromuscular adaptation on BDFT needs to be accounted for. Hence, a sensible requirement that follows from this is that a mitigation method should be experimentally tested for a range of different neuromuscular settings, including compliant, stiff and relaxed settings.

9.3   Remaining challenges

In the previous sections the methods, results and conclusions contained in this thesis were discussed. On some occasions it was indicated how one could proceed to further our understanding, but always by taking a result obtained within the scope of this thesis as a starting point. This section aims to augment the previous sections by addressing some of the questions that were not at all, or not satisfactorily addressed so far.

9.3.1 Biodynamic feedthrough in actual vehicles

In the context of this thesis, BDFT was studied in the laboratory only. As a result, the severity of the actual problem of BDFT occurring in, e.g., helicopters and excavators, was not thoroughly addressed. Due to the fact that BDFT is relatively unknown it is possible that BDFT occurs more frequently than is currently assumed. An investigation into practical real-world BDFT occurrences may reveal that many BDFT issues are not recognized as such, are accepted 'as is', or remedied through suboptimal methods.

It is known that BDFT is related to some vehicle-specific phenomena such as Aircraft / Rotorcraft-Pilot Couplings (A/RPCs) or roll ratcheting. It would be interesting to see how these phenomena are exactly influenced by BDFT and how knowledge of one phenomenon can benefit the understanding of the other.

Using techniques proposed in this thesis it is possible to address questions such as how RPCs may be triggered by BDFT induced inputs. By combining a BDFT model, obtained for a representative cockpit environment, with an accurate rotorcraft model and pilot model, one could investigate the possible contribution of BDFT into adverse couplings (see, for example, Hess [2010]). Such research efforts would establish valuable connections between the fundamental BDFT work that has been done in the laboratory, and the adverse couplings occurring in actual vehicles.

9.3.2 The effect of cognitive corrective control

It is known that an operator is capable of applying cognitive corrective control actions, counteracting the involuntary inputs induced by vehicle accelerations and thus reducing the effects of BDFT. This ability, and especially its limitations, are not well understood and deserve more attention. It is possible that the susceptibility of a vehicle to BDFT effects is 'masked' by continuous low-level skill-based corrective control actions, which leaves us unaware of the existence of a problem.

The fact that an operator is able to cancel BDFT cognitively does not mean the mitigation comes 'for free'. Even low-level skill-based corrective control actions impose a load on the human operator. Effective BDFT mitigation approaches may allow an operator to devote more of their mental or physical resources to vehicle control, making operations easier and safer.

9.3.3   The effect of preview

Knowledge on future exposure to acceleration disturbances (preview) allows a human operator to exploit higher-order cognitive capabilities to deal with BDFT ('brace/relax for impact'). Such a strategy can, for example, be found in mountain biking: the experienced mountain biker adapts the setting of his/her neuromuscular system to the terrain ahead. The likely advantage of these preview-based adaptations is that they are less costly than post-hoc cognitive corrective control actions. Investigating whether and how preview helps human operators to deal with BDFT is highly relevant in understanding human's ability to deal with BDFT effects in practice.

9.3.4   Model-based force cancellation

In this thesis, model-based *signal cancellation* was experimentally tested. An interesting alternative to this is *force cancellation* [Repperger, 1995; Sövényi and Gillespie, 2007]. An important difference between the two types is that force cancellation adheres to the shared control paradigm [Griffiths and Gillespie, 2004], providing feedback at the control device to the human operator about the activity of the controller. This in contrast to the input mixing approach [Abbink and Mulder, 2010] used in signal cancellation, where the canceling signal is applied after the control device input. The expected benefit of the force cancellation approach is that it improves the awareness of the operator on the controller's activity, keeping the human 'in the loop'. A possible disadvantage is that it is unknown whether and how the human operator will respond to these additional forces. A human operator may adjust his/her neuromuscular dynamics in response, which may lead to adaptations of the magnitude of the canceling force, causing another adaptation of the human, etc. This could possibly cause an oscillatory cycle of adaptations where the human and controller mutually adapt to each other. Further research into the force cancellation approach and comparisons with signal cancellation are very relevant future research directions.

9.3.5 Operator state observation

Throughout this thesis, it has been stated several times that mitigating BDFT effectively requires knowledge regarding the current human operator neuromuscular state. Obtaining this knowledge in an online fashion requires new techniques for fast and reliable online system identification and parameter estimation. It should be noted that in such an approach one does not necessarily need to rely on traditional admittance estimates. For adaptive BDFT mitigation it suffices to obtain a metric that can be measured online, which reliably correlates to the BDFT dynamics, allowing for adaptation of the BDFT mitigation system.

Possible candidates for these metrics are grip force or forearm EMG signals. Such metrics would provide information on the neuromuscular 'state' of the human operator. Operator state observation is a topic that reaches far beyond the field of BDFT research alone. It currently receives well-deserved attention from different disciplines within the human-machine interaction community. Obtaining a successful estimate of the operator's mental or neuromuscular state may inaugurate a new chapter in human-machine interaction, one where the machine adapts to the needs of the human instead of the other way around.

9.4 A review of the research goal

The research goal was to *increase the understanding of BDFT to allow for effective and efficient mitigation of the BDFT problem.*

With respect to *increasing the understanding* it can be stated that considerable progress has been made in measuring, analyzing and modeling biodynamic feedthrough. Particularly regarding the role of the neuromuscular system, which proved to be a highly relevant influence on the BDFT dynamics, many new insights were obtained. In addition to that, the results presented in this thesis also showed how BDFT is composed of different dynamical relationships and that understanding these allows for understanding the influence of the control device dynamics or the presence of an armrest. The most important fields where we didn't increase our understanding are the practical occurrence of biodynamic feedthrough in actual vehicles and the complexities of closed-loop BDFT effects. These remain highly relevant topics for future studies.

With respect to *mitigation of the BDFT problem*, progress was made in two main directions. The first is in the area of identifying the available mitigation approaches with their strengths and weaknesses. This provides us with a clear overview as to which options we have at our disposal. The second direction is the practical investigations of two specific mitigation approaches: the often overlooked armrest and the novel method of admittance-adaptive model-based signal cancellation.

Both mitigation methods showed to be successful in mitigating biodynamic feedthrough, so both methods can be regarded as *effective*. When it comes to the question whether they were *efficient*, opinions may differ. The armrest is surely the more efficient of the two, requiring nothing else than the installation of a simple piece of hardware at the operator's control station. The admittance-adaptive model-based signal cancellation approach can be regarded as more efficient than a force cancellation technique, as the latter requires the use of an 'active stick' that is capable of generating forces. Also, the mathematical model that was used can be regarded as more efficient than a physical model, due to the latter's higher complexity. However, the fact that the model-based approach requires an individual BDFT model that should be adapted to match the current neuromuscular setting of the operator makes the approach far from straightforward to implement. More research will be required in order to reduce the complexities, and so increase the efficiency, of this approach.

CHAPTER 10

Conclusions and recommendations

10.1 Introduction

Biodynamic feedthrough (BDFT) is defined in this thesis as follows:

Biodynamic feedthrough

The transfer of accelerations through the human body during the execution of a manual control task, causing involuntary forces being applied to the control device which may result in involuntary control device deflections.

The research goal of the work presented here was:

Research goal:

Increase the understanding of BDFT to allow for effective and efficient mitigation of the BDFT problem.

This thesis focused particularly on *the influence of the variable neuro-muscular dynamics on BDFT dynamics*. The approach of the research work consisted of three parts: first, a method was developed to accurately measure BDFT. Then, several BDFT models were developed to increase our understanding of how this phenomenon can be captured and described. Finally, using the insights from the previous steps, a novel approach to BDFT mitigation was proposed.

In the following, a concise reiteration of the conclusions obtained throughout the different chapters is presented in Sections 10.2–10.5. This is followed by a synthesis into the general conclusions of this thesis in Section 10.6. Finally, in Section 10.7, some recommendations for future research are presented.

10.2   Measuring biodynamic feedthrough

One of the main challenges in BDFT research is to understand its dependency on the dynamics of the human body and thus, indirectly, on the behavior of the human operator. In order to gain a proper understanding of the dependency of BDFT dynamics on neuromuscular dynamics, both need to be measured simultaneously. A measurement method that allows for this was proposed in **Chapter 2**.

Overview of Chapter 2   Neuromuscular admittance is a dynamic property of a limb, characterized by the relationship between force input and position output of a limb. By simultaneously measuring BDFT dynamics and admittance, valuable insights can be gained on how the variable neuromuscular properties affect BDFT. A measurement method that allows for this was experimentally validated. In the experiments the admittance was varied using the classical tasks: a position task (PT) or 'stiff task', a force task (FT) or 'compliant task', and a relax task (RT). By following the PT, FT and RT task instructions the subject attains, respectively, a maximally stiff setting of the neuromuscular system (a low admittance), a maximally compliant setting (a high admittance) and a passive setting.

Conclusions of Chapter 2   It was concluded that the proposed method was successful in the simultaneous measurement of admittance and biodynamic feedthrough. The results for the admittance measurements are comparable to results found in other studies. High coherence values were found, which signify good signal-to-noise ratios, indicating that the admittance estimates are indeed reliable. Also for the BDFT measurements high coherences were found. Furthermore, the BDFT dynamics that were obtained between subjects show comparable shapes and features.

> **Main finding of Chapter 2:**
>
> Neuromuscular dynamics have a strong influence on BDFT dynamics.

10.3 Analyzing biodynamic feedthrough

There is little consensus on how to approach biodynamic feedthrough problems in terms of definitions, nomenclature and mathematical descriptions. The abundance of and diversity between names used in the literature referring to the same or similar phenomena impedes communication between researchers and comparison between studies. In order to improve this situation a framework for biodynamic feedthrough was proposed in **Chapter 3**.

Overview of Chapter 3 The framework for biodynamic feedthrough analysis aims to provide a common ground to study, discuss and understand BDFT and its related problems. Using this framework, old and new BDFT research can be (re)interpreted, evaluated and compared. Also, and equally important, the framework itself allows for gaining new insights into the BDFT phenomenon. Different types of BDFT dynamics were defined and mathematical relationships were derived that describe how these dynamics relate to each other. The proposed relationships were validated using experimental data.

Conclusions of Chapter 3 A distinction was made between the effects of BDFT on the generation of involuntary control forces and on the generation of involuntary control device deflections (positions). It was proposed to label them BDFT to forces (B2F) and BDFT to positions (B2P) respectively. In addition to B2P, which is the focus of most existing BDFT literature, B2F provides valuable insights. The B2F dynamics can be defined in two different ways, giving rise to the terms biodynamic feedthrough to forces in open-loop (B2FOL) and biodynamic feedthrough to forces in closed-loop (B2FCL).

The mathematical relationships were validated. It was shown that they provide additional insights, such as how to accurately predict the effect of control device dynamics on BDFT dynamics. Furthermore, by applying the framework it was shown that the approach and results put forward in three selected studies could be successfully reinterpreted.

> **Main finding of Chapter 3:**
>
> The framework proved to be useful in both interpreting previous BDFT studies and in gaining new insights.

10.4   Modeling biodynamic feedthrough

The currently existing BDFT models can be roughly divided in two groups: physical BDFT models and black box BDFT models. Physical models are geared towards providing a physical representation of the BDFT phenomenon, using a-priori knowledge and physical principles. A drawback of physical models is that they are often complex, which makes implementation and parameter estimation of these models challenging. Black box models, in contrast, aim to provide an efficient BDFT description at 'end-point level'. The black box modeling approach does not rely on physical principles and the resulting models are therefore often easier to use compared to their physical counterparts. A drawback of black box models is that they do not provide the same level of insight as physical models do. In this thesis, two novel BDFT models are proposed. The first is a physical model, described in **Chapter 4**.

Overview of Chapter 4   The physical BDFT model serves primarily the purpose of increasing the understanding of the relationship between admittance and BDFT. The model was constructed by extending a well-known and well-documented admittance model. The challenging process of parameter estimation was handled by using a two-stage parameter estimation approach. First, the parameters of the admittance model were estimated, using established techniques. Then, these parameter values were used in the BDFT model and the remaining parameters were estimated. The quality of the model was evaluated in frequency and time domain.

Conclusions of Chapter 4 The results provide strong evidence that the proposed physical BDFT model, including the proposed method of estimating its parameters, allows for accurate BDFT modeling across different subjects and across control tasks. The model parameters can be used to gain more insights regarding the physical principles of BDFT. As an additional validation, it was confirmed that the influence of the control device dynamics is correctly represented by the model.

Main finding of Chapter 4:

The physical BDFT model provides an accurate physical description of the BDFT dynamics, increasing our fundamental understanding of the BDFT phenomenon.

The mathematical model was described in **Chapter 5**.

Overview of Chapter 5 The mathematical BDFT model aims to fill the gap between currently existing black box models and physical models. The model structure was obtained through asymptote modeling, which offers a structural method to design a model's transfer function. Asymptote modeling relies on tuning the asymptotic behavior of base functions and combining these to create complexer model structures. The method was used here to model the B2FCL dynamics measured in a rotorcraft setup. The resulting model was thoroughly evaluated in both frequency and time domain. Furthermore, the model's performance was compared to two black box models and to the physical model proposed in Chapter 4.

Conclusions of Chapter 5 The results show that the mathematical BDFT model provides a highly accurate description of BDFT dynamics. Using the proposed parameter values the model can be directly implemented in many typical rotorcraft BDFT studies. The 'global scope model', obtained by averaging the results of all subjects, performs well, although 'individual scope models' are slightly superior. The comparison between the mathematical BDFT model and other models showed that the proposed mathematical model outperforms two black box models. The performance of the physical model and the mathematical model proved to be comparable, but the parameter estimation procedure for the mathematical model is considerably easier.

The accuracy of the mathematical BDFT model is evidence that the asymptote modeling approach was successful. The method is likely to be useful in other modeling problems as well.

> **Main findings of Chapter 5:**
>
> The mathematical BDFT model is highly accurate, outperforms several black box models and is easier in use than the physical model.
> Asymptote modeling proved successful in obtaining an accurate and versatile model structure for the mathematical BDFT model.

10.5 Mitigating biodynamic feedthrough

Several BDFT mitigation techniques were discussed in **Chapter 6**.

Overview of Chapter 6 Using the BDFT system model, the available BDFT mitigation techniques were discussed. In total, seven different solution types, each providing one or more solution approaches, were identified. After discarding the solution types that do not meet predefined requirements, two solution types remained that were deemed most promising. Measures of the first solution type – passive support/restraining systems (e.g., seat belts and armrests) – are already commonly applied. Studies have shown that these are not sufficient to remove BDFT completely. The second solution type is model-based BDFT cancellation, where use is made of a BDFT model to compute a canceling signal. This approach has received some attention in the literature, but only very few experimental implementations have been tested. The potential of signal cancellation was further investigated using models with different levels of generality, ranging from 'run level' (specific) to 'global level' (general).

Conclusions of Chapter 6   The results show that subject level models applied to run level data provided between 60% to 80% cancellation of the BDFT effects. When using global level models the cancellation reduced to a value between 40% and 50%. This implies that averaging over the subjects reduces the applicability of the BDFT model for BDFT cancellation. Signal cancellation with global level models is possible, but they are outperformed by subject level models. When attempting signal cancellation across tasks it is observed that good cancellation can only be achieved when the model matches the task. When compared to the results for the generality levels, it showed that adaptation to task, i.e., to the neuromuscular dynamics of the human operator, is of particular importance. A failure to identify changes in the neuromuscular settings of the human operator and adapting the model accordingly leads to suboptimal or incorrect control actions.

> **Main finding of Chapter 6:**
>
> Signal cancellation is only a promising mitigation method for biodynamic feedthrough problems if the model can be adapted to both the subject and, particularly, the task.

The effectiveness of an armrest in mitigating biodynamic feedthrough was investigated in **Chapter 7**.

Overview of Chapter 7   The BDFT dynamics were measured with and without the armrest for the three classical tasks (PT, RT, FT). The effectiveness of the armrest was evaluated through the ratio function (RF) which computes the ratio between the dynamics measured with and without the armrest. The RF was computed for B2P, B2FCL and B2FOL dynamics. To investigate the effect of the armrest on the human body dynamics the RF was also calculated for the admittance and force disturbance feedthrough (FDFT) dynamics.

Conclusions of Chapter 7 The results show that, for the three levels of neuromuscular admittance investigated, the presence of an armrest decreases the level of both B2P and B2FOL. This implies that both the involuntary deflections of the control device and involuntary forces applied to the control device are reduced by the addition of an armrest. The results furthermore provide the novel insight that the effect of the armrest varies strongly with disturbance frequency and neuromuscular admittance. Only a minor effect of the addition of the armrest was observed on the response to force disturbances, i.e., the admittance and FDFT dynamics. This enforces the conclusion that all differences in the BDFT dynamics are due to the presence of the armrest and not due to changes in the human body dynamics. The analysis of the B2FCL dynamics showed that these dynamics are not well suited to evaluate a mitigation approach, as they lack a clear interpretation. This is not the case for the admittance, FDFT, B2P and B2FOL dynamics.

Main finding of Chapter 7:

An armrest is an effective tool in mitigating biodynamic feedthrough. The effectiveness depends on the disturbance frequency and neuromuscular admittance.

The results obtained by a novel approach to BDFT mitigation were provided in **Chapter 8**.

Overview of Chapter 8 An admittance-adaptive model-based signal cancellation approach was proposed. What differentiates this BDFT mitigation method from other approaches is that it accounts for adaptations in the neuromuscular dynamics of the human body. The approach was tested, as proof-of-concept, in an experimental setup where subjects inside a motion simulator were asked to fly a simulated vehicle through a virtual tunnel. The cancellation was based on a BDFT model that was used to compute the involuntary control inputs. In the experiment there were two independent variables: the control task (PT or RT), and the condition, for which there were four levels: the static condition (STA), in which no acceleration disturbances were applied; the motion condition (MOT), in which acceleration disturbances were applied but cancellation was inactive; the cancellation condition (CAN), in which acceleration disturbances were applied and the cancellation was active. In order to investigate the importance of the influence of neuromuscular setting on cancellation, a fourth condition was added: the incongruent condition (INC), in which the cancellation was done in an 'incongruent' fashion, i.e., the PT model was applied in the RT sections and vice-versa.

Conclusions of Chapter 8 From the results of the experiment it was concluded that the BDFT model described the measured BDFT dynamics well. The performance metrics provided congruent results regarding the effect of the motion disturbance and cancellation on control performance and control effort. From the results it was concluded that BDFT occurred in the motion condition (MOT), decreasing control performance and increasing effort with respect to the static condition (STA). Furthermore, with cancellation active (CAN) the performance and effort were largely restored to the values obtained in the static condition (STA). And finally it was observed that incongruent cancellation (INC) leads to lower performance and higher effort than obtained with congruent cancellation (CAN), but only for the PT. The absence of an effect for the RT case can be explained by the fact that some subjects behaved differently in the identification measurements and the actual cancellation experiment, making both the PT and RT model equally applicable.

> **Main finding of Chapter 8:**
>
> The admittance-adaptive model-based signal cancellation approach was successful and largely removed the negative effects of BDFT on the control performance and control effort.

10.6   General conclusions

By synthesizing the results that were obtained throughout the individual chapters, the following general conclusions can be drawn:

A fragmented research environment: Biodynamic feedthrough is a complex process, in which many and often still poorly understood factors play a role. This limited understanding hampers the development of practical and generally applicable solutions. The diversity in methods and terminology in existing literature impedes a clear communication between researchers. This has resulted in a fragmented research environment where BDFT problems are investigated on a case-to-case basis. An increased consensus in the definitions, nomenclature and mathematical descriptions would benefit the understanding of biodynamic feedthrough, improve the communication between researchers and facilitate the comparison between studies.

The influence of human body dynamics: Neuromuscular dynamics, and especially the variability thereof, have an important influence that needs to be accounted for when measuring, modeling and mitigating biodynamic feedthrough effects. Biodynamic feedthrough dynamics vary both between different persons, as well as within one person over time. When measuring biodynamic feedthrough, it is important to measure, monitor and/or control the neuromuscular dynamics of the subjects. When modeling biodynamic feedthrough, the models should be able to cope with between-subject and within-subject variability. When mitigating biodynamic feedthrough, the effectiveness of simple mitigation strategies are likely to vary depending on the human body dynamics. Only some mitigation strategies allow for taking this variability into account.

Modeling biodynamic feedthrough: There are many possible ways in which biodynamic feedthrough can be modeled. The preferred way strongly depends on the intended use of the model. Individual BDFT models (for one person) are often superior to generalized BDFT models (for a group of persons).
The community of researchers that is interested in *using* BDFT models is much larger than the community of researchers that is *developing* those models. Therefore, biodynamic feedthrough models should be designed such that specialists can incorporate novel insights with the necessary degree of detail, while retaining sufficient practical usability to allow the result to be used by a larger user community.

Mitigating biodynamic feedthrough: Many different ways of mitigating biodynamic feedthrough exist. Which of these is the optimal approach for a given situation depends on many different factors, such as the efficiency and the effectiveness of an approach. A simple solution like the armrest has shown to be both highly efficient and effective.
The mitigation of biodynamic feedthrough through model-based signal cancellation is a powerful and versatile approach, but only successful if the biodynamic feedthrough model is adapted to both human operator and control task. There are several obstacles to overcome before this method can be put to practical use in actual vehicles, mainly regarding the adaption of the model parameters in response to changes in the neuromuscular dynamics of the human operator. Research efforts directed at addressing this challenge should be encouraged.

10.7   Recommendations

Based on the discussion of the results in Chapter 9, the following recommendations can be made for future research.

- Novel identification methods are required that allow for the measurement of BDFT dynamics in more natural control tasks in which the settings of the neuromuscular system are closer to those encountered in actual vehicle control. Preferably, these new methods should not rely on the limiting assumption of linearity and time-invariance of the dynamics. They should provide a (close to) online estimate of the BDFT dynamics and/or neuromuscular dynamics of the human operator. Such methods are a requirement for the successful implementation of adaptive model-based BDFT mitigation techniques.

- The further use of the framework proposed in this thesis should be encouraged. The fragmentation in methods and terminology that currently characterizes the BDFT research field is otherwise likely to remain. Furthermore, the framework should be improved and extended based on future insights regarding, e.g., closed-loop BDFT effects.

- Future BDFT modeling efforts should be directed at obtaining models of limited complexity, without compromising fidelity. The mathematical BDFT model, proposed in this thesis, seems to possess these traits. The mathematical BDFT model should be put to the test in several different BDFT problems to investigate its value. The asymptote modeling technique, used to construct the mathematical model, should be applied to other modeling problems to refine the methodology.

- The laboratory-based research regarding BDFT has been fruitful and will remain useful in future BDFT studies. More practical research, however, directed at the occurrence of BDFT in actual vehicles and at establishing connections between the laboratory work and real-world problems, is indispensable in order to find effective solutions to the BDFT problems occurring around us.

Bibliography

Abbink, D. A., *Neuromuscular analysis of haptic gas pedal feedback during car following*, Ph.D. thesis, TU Delft, 2006.

Abbink, D. A. and Mulder, M., "Neuromuscular analysis as a guideline in designing shared control," Zadeh, M. H. (editor), *Advances in Haptics*, chap. 27, pp. 499–516, INTECH, Austria, Apr. 2010.

Abbink, D. A., Mulder, M., van der Helm, F. C. T., Mulder, M., and Boer, E. R., "Measuring neuromuscular control dynamics during car following with continuous haptic feedback," *IEEE Trans. on Systems, Man, and Cybernetics, Part B: Cybernetics*, vol. 41, no. 5, pp. 1239–1249, 2011, doi:10.1109/TSMCB.2011.2120606.

Allen, R. E., Jex, H. R., and Magdaleno, R. E., "Manual control performance and dynamics response during sinusoidal vibration," Tech. Rep. AD-773844, System Technology Inc., Hawthorne (CA), Oct. 1973.

Anon., "Apollo terminology," Tech. Rep. NASA- SP-6001, NASA, Aug. 1963.

Arai, F., Tateishi, J., and Fukuda, T., "Dynamical analysis and suppression of human hunting in the excavator operation," *9th IEEE Int. Workshop on Robot and Human Interactive Communication, RO-MAN, Osaka, Japan*, pp. 394–399, 2000, doi:10.1109/ROMAN.2000.892530.

Banerjee, D., Jordan, L. M., and Rosen, M. J., "Modeling the effects of inertial reactions on occupants of moving power wheelchairs," *Rehab. Eng. and Assistive Tech. Soc. of N. America Conf. (RESNA), Salt Lake City (UT)*, pp. 220–222, Jun. 1996.

Bennet, L., "Powered wheelchair bucking," *J. of rehab. research and development*, vol. 24, no. 2, pp. 81–86, 1987.

Brown, J. L. and Lechner, M., "Acceleration and human performance: a survey of research," *J. of Aviation Medicine*, vol. 27, no. 1, pp. 32–49, Feb. 1956.

Brown, M. C., Engberg, I., and Matthews, P. B. C., "The relative sensitivity to vibration of muscle receptors of the cat," *J. of Physiology*, vol. 192, no. 1, pp. 773–800, 1967.

Cathers, I., O'Dwyer, N., and Neilson, P., "Dependence of stretch reflexes on amplitude and bandwidth of stretch in human wrist muscle," *Experimental Brain Research*, vol. 129, no. 2, pp. 278–287, Nov. 1999.

Damveld, H. J., Abbink, D. A., Mulder, M., Mulder, M., van Paassen, M. M., van der Helm, F. C. T., and Hosman, R. J. A. W., "Measuring the contribution of the neuromuscular system during a pitch control task," *AIAA Modeling and Simulation Technologies Conf., Chicago (IL)*, Aug. 2009.

Damveld, H. J., Abbink, D. A., Mulder, M., Mulder, M., van Paassen, M. M., van der Helm, F. C. T., and Hosman, R. J. A. W., "Identification of the feedback component of the neuromuscular system in a pitch contol task," *AIAA Guidance, Navigation, and Control Conf., Toronto, Canada*, Aug. 2010.

de Vlugt, E., *Identification of spinal reflexes*, Ph.D. thesis, TU Delft, 2004.

de Vlugt, E., Schouten, A. C., and van der Helm, F. C. T., "Quantification of intrinsic and reflexive properties during multijoint arm posture," *J. of Neuroscience Methods*, vol. 155, no. 2, pp. 328–349, Sep. 2006, doi:10.1016/j.jneumeth.2006.01.022.

Dieterich, O., Götz, J., DangVu, B., Haverdings, H., Masarati, P., Pavel, M. D., Jump, M., and Gennaretti, M., "Adverse rotorcraft-pilot coupling: Recent research activities in europe," *34th Eur. Rotorcraft Forum, Liverpool, UK*, pp. 1–20, Sep. 2008.

Dimasi, F. P., Allen, R. E., and Calcaterra, P. C., "Effect of vertical active vibration isolation on tracking performance and on ride qualities," Tech. Rep. NASA CR-2146, NASA, Nov. 1972.

Donati, P. M. and Bonthoux, C., "Biodynamic response of the human body in the sitting position when subjected to vertical vibration," *J. of Sound and Vibration*, vol. 90, no. 3, pp. 432–442, 1983.

Gabel, R. and Wilson, G. J., "Test approaches to external sling load instabilities," *J. of the American Helicopter Society*, vol. 13, no. 3, pp. 44–54, 1968, doi:10.4050/JAHS.13.44.

Gennaretti, M., Serafini, J., Masarati, P., and Quaranta, G., "Effects of biodynamic feedthrough in rotorcraft/pilot coupling: collective bounce case," *J. of Guidance, Control & Dynamics*, pp. 1–13, 2013, doi:10.2514/1.61355.

Gillespie, R. B., C. Hasser, C., and Tang, P., "Cancellation of feedthrough dynamics using a force reflecting joystick," *ASME Dynamic Systems and Control Conf., Hollywood (CA)*, pp. 319–326, Oct. 1999.

Griffin, M. J., "The evaluation of vehicle vibration and seats," *Applied Ergonomics*, vol. 9, no. 1, pp. 15–21, 1978, doi:10.1016/0003-6870(78)90214-4.

Griffin, M. J., "Biodynamic response to whole-body vibration," *The Shock and Vibration Digest*, vol. 13, no. 3, pp. 2–12, 1981, doi:10.1177/058310248101300803.

Griffin, M. J., *Handbook of human vibration*, Academic Press, 1st edn., 1990.

Griffin, M. J., "The validation of biodynamic models," *Clinical Biomechanics*, vol. 16, no. 1, pp. S81–S92, 2001, doi:10.1016/S0268-0033(00)00101-7.

Griffin, M. J. and Whitham, E. M., "Individual variability and its effect on subjective and biodynamic response to whole-body vibration," *J. of Sound and Vibration*, vol. 58, no. 2, pp. 239–250, 1978, doi:10.1016/S0022-460X(78)80078-9.

Griffiths, P. and Gillespie, R. B., "Shared control between human and machine: haptic display of automation during manual control of vehicle heading," *12th Int. Symposium on Haptic Interfaces for Virtual Environment and Teleoperator Systems, HAPTICS '04*, pp. 358–366, Mar. 2004, doi:10.1109/HAPTIC.2004.1287222.

Hamel, P. G., "Rotorcraft-pilot coupling: A critical issue for highly augmented helicopters?" *AGARD Flight Vehicle Integration Panel Symposium on 'Advances in Rotorcraft Technology', Ottawa, Canada*, pp. 21–1–21–9, May 1996.

Hess, R. A., "Theory for roll-ratchet phenomenon in high-performance aircraft," *J. of Guidance, Control & Dynamics*, vol. 21, no. 1, pp. 101–108, 1998, doi:10.2514/2.4203.

Hess, R. A., "Modeling biodynamic interference in helicopter piloting tasks," *AHS Aeromechanics Specialists Conf., San Francisco (CA)*, pp. 496–504, Jan. 2010.

Humphreys, H., Book, W., and Huggins, J., "Compensation for biodynamic feedthrough in backhoe operation by cab vibration control," *IEEE Int. Conf. on Robotics and Automation*, pp. 4284–4290, May 2011, doi:10.1109/ICRA.2011.5979808.

Humphreys, H. C., Book, W. J., and Feigh, K. M., "Development of controller-based compensation for biodynamic feedthrough in a backhoe," *Proc. of the Institution of Mechanical Engineers, Part I: J. of Systems and Control Engineering*, vol. 228, no. 2, pp. 107–120, 2014, doi:10.1177/0959651813506188.

Idan, M. and Merhav, S. J., "Effects of biodynamic coupling on the human operator model," *J. of Guidance, Control & Dynamics*, vol. 13, no. 4, pp. 630–637, 1990, doi:10.2514/2.4203.

Jewell, W. F. and Citurs, K. D., "Quantification of cross-coupling and motion feedthrough for multiaxis controllers used in an air combat flying task," *20th Annual Conf. on Manual Control*, vol. 1, pp. 79–90, 1984.

Jex, H. R., "Problems in modeling man-machine control behavior in biodynamic environments," *7th Annual Conf. on Manual Control*, pp. 3–13, 1972.

Jex, H. R. and Magdaleno, R. E., "Biomechanical models for vibration feedthrough to hands and head for a semisuspine pilot," *Aviation, Space, and Environmental Medicine*, vol. 49, no. 1, pp. 304–316, 1978.

Jex, H. R. and Magdaleno, R. E., "Modeling biodynamic effects of vibration," Tech. Rep. ADA-073819 / STI-1037-1, Systems Technology Inc., 1979.

Jones, W. L., "Biodynamics bibliography (1966-1969)," Tech. Rep. NASA-TM-X-67138, NASA, 1971.

Katzourakis, D. I., Abbink, D. A., Velenis, E., Holweg, E., and Happee, R., "Driver's arms' time-variant neuromuscular admittance during real car test-track driving," *IEEE Trans. on Instrumentation and Measurement*, vol. PP, no. 99, pp. 1–1, 2013, doi: 10.1109/TIM.2013.2277610.

Larue, M., "The effects of vibration on accuracy of a positioning task," *J. of Environmental Science*, vol. 8, no. 1, pp. 33–35, 1965.

Lasschuit, J., Lam, T. M., Mulder, M., van Paassen, M. M., and Abbink, D. A., "Measuring and modeling neuromuscular system dynamics for haptic interface design," *AIAA Modeling and Simulation Technologies Conf., Honululu, Hawaii*, pp. 304–316, Aug. 2008.

Lewis, C. H., *The development of a predictive model of the effects of vibration on human operator performance*, Master's thesis, University of Southampton, 1974.

Lewis, C. H. and Griffin, M. J., "The effects of vibration on manual control performance," *Ergonomics*, vol. 19, no. 2, pp. 203–216, 1976, doi:10.1080/00140137608931532.

Lewis, C. H. and Griffin, M. J., "The interaction of control gain and vibration with continuous manual control performance," *J. of Sound and Vibration*, vol. 55, no. 4, pp. 553–562, 1977, doi:10.1016/S0022-460X(77)81179-6.

Lewis, C. H. and Griffin, M. J., "Predicting the effects of dual-frequency vertical vibration on continuous manual control performance," *Ergonomics*, vol. 21, no. 8, pp. 637–650, 1978a, doi:10.1080/00140137808931765.

Lewis, C. H. and Griffin, M. J., "A review of the effects of vibration on visual acuity and continuous manual control, part i: Visual acuity," *J. of Sound and Vibration*, vol. 56, no. 3, pp. 383–413, 1978b, doi:10.1016/S0022-460X(78)80155-2.

Lewis, C. H. and Griffin, M. J., "A review of the effects of vibration on visual acuity and continuous manual control, part II: Continuous manual control," *J. of Sound and Vibration*, vol. 56, no. 3, pp. 415–457, 1978c, doi:10.1016/S0022-460X(78)80156-4.

Lewis, C. H. and Griffin, M. J., "Mechanisms of the effects of vibration frequency, level and duration on continuous manual control performance," *Ergonomics*, vol. 22, no. 7, pp. 855–889, 1979, doi:10.1080/00140137908924662.

Ljung, L., *System identification-Theory for the user*, Prentice Hall, Inc., 2nd edn., 1999.

Lovesey, E. J., "An investigation into the effects of dual axis vibration, restraining harness, visual feedback and control force on a manual positioning task," Tech. Rep. TR-71213, RAE, 1971a.

Lovesey, E. J., "Some effects of dual-axis heave and sway vibrations on compensatory tracking," Tech. Rep. EP-484, RAE, 1971b.

Maddan, S., Walker, J. T., and Miller, J. M., "Does size really matter? A reexamination of Sheldon's somatotypes and criminal behavior," *The Social Science Journal*, vol. 45, no. 2, pp. 330–344, 2008, doi:10.1016/j.soscij.2008.03.009.

Masarati, P., Quaranta, G., and Jump, M., "Experimental and numerical helicopter pilot characterization for aeroelastic rotorcraft-pilot couplings analysis," *Proc. of the Institution of Mechanical Engineers, Part G: J. of Aerospace Engineering*, vol. 227, no. 1, pp. 125–141, Jan. 2013, doi:10.1177/0954410011427662.

Masarati, P., Quaranta, G., Lu, L., and Jump, M., "A closed loop experiment of collective bounce aeroelastic Rotorcraft-Pilot Coupling," *J. of Sound and Vibration*, vol. 333, no. 1, pp. 307–325, 2014, doi:10.1016/j.jsv.2013.09.020.

Masarati, P., Quaranta, G., Serafini, J., and Gennaretti, M., "Numerical investigation of aeroservoelastic rotorcraft-pilot coupling," *AIDAA 2008 XIX Congresso Nazionale AIDAA, Forlì, Italy*, Sep. 2007.

Mattaboni, M., Fumagalli, A., Jump, M., Masarati, P., and Quaranta, G., "Biomechanical pilot properties identification by inverse kinematics/inverse dynamics multibody analysis," *ICAS 26th Congress of the Int. Council of the Aeronautical Sciences, Anchorage (AK)*, Sep. 2008.

Mayo, J. R., "The involuntary participation of a human pilot in a helicopter collective control loop," *15th Eur. Rotorcraft Forum, Amsterdam, the Netherlands*, pp. 81–001–81–012, Sep. 1989.

McLeod, R. W. and Griffin, M. J., "Performance of a complex manual control task during exposure to vertical whole-body vibration between 0.5 and 5.0 Hz," *Ergonomics*, vol. 31, no. 8, pp. 1193–1203, 1988, doi:10.1080/00140138808966757.

McLeod, R. W. and Griffin, M. J., "Review of the effects of translational whole-body vibration on continuous manual control performance," *J. of Sound and Vibration*, vol. 133, no. 1, pp. 55–115, 1989, doi:10.1016/0022-460X(89)90985-1.

McLeod, R. W. and Griffin, M. J., "Effects of duration and vibration on performance of a continuous manual control task," *Ergonomics*, vol. 36, no. 6, pp. 645–659, 1993, doi:10.1080/00140139308967926.

McLeod, R. W. and Griffin, M. J., "Mechanisms of vibration-induced interference with manual control performance," *Ergonomics*, vol. 38, no. 7, pp. 1431–1444, 1995, doi:10.1080/00140139508925200.

McRuer, D. T. and Jex, H. R., "A review of quasi-linear pilot models," *IEEE Trans. on Human Factors in Electronics*, vol. 8, no. 3, pp. 231–249, Sep. 1967.

Mitchell, D., Hoh, R., Adolph, A. J., and Key, D., "Ground based simulation evaluation of the effects of time delays and motion on rotorcraft handling qualities," Tech. Rep. USAAVSCOM TR 91-A-010, AD-A256 921, Systems Technology Inc., Jan. 1992.

Mugge, W., Abbink, D. A., Schouten, A., Dewald, J., and van der Helm, F. C. T., "A rigorous model of reflex function indicates that position and force feedback are flexibly tuned to position and force tasks," *Experimental Brain Research*, vol. 200, no. 3, pp. 325–340, 2009, doi:10.1007/s00221-009-1985-0.

Mugge, W., Abbink, D. A., and van der Helm, F. C. T., "Reduced power method: how to evoke low-bandwidth behaviour while estimating full-bandwidth dynamics," *IEEE 10th Int. Conf. on Rehabilitation Robotics. ICORR, Noordwijk aan Zee, The Netherlands*, pp. 575–581, Jun. 2007.

Mulder, M. and Mulder, J. A., "Cybernetic analysis of perspective flight-path display dimensions," *J. of Guidance, Control & Dynamics*, vol. 28, no. 3, pp. 398–411, 2005, doi:10.2514/1.6646.

Mulder, M., Verspecht, T., Abbink, D. A., van Paassen, M. M., Balderas, D. C., Schouten, A., de Vlugt, E., and Mulder, M., "Identification of time-variant neuromuscular admittance using wavelets," *IEEE Int. Conf. on Systems, Man, and Cybernetics, Anchorage (AK)*, pp. 1474 –1480, Oct. 2011, doi:10.1109/ICSMC.2011.6083879.

Nakamura, H., Abbink, D. A., and Mulder, M., "Is grip strength related to neuromuscular admittance during steering wheel control?" *IEEE Int. Conf. on Systems, Man, and Cybernetics, Anchorage (AK)*, pp. 1658–1663, 2011, doi:10.1109/ICSMC.2011.6083909.

National Research Council, *Aviation Safety and Pilot Control: Understanding and Preventing Unfavorable Pilot-Vehicle Interactions*, National Academies Press, Washington, D.C., 1997, ISBN 9780309056885.

Pavel, M. D., Malecki, J., DangVu, B., Masarati, P., Gennaretti, M., Jump, M., Jones, M., Smaili, H., Ionita, A., and Zaicek, L., "Present and future trends in rotorcraft pilot couplings (RPCs): A retrospective survey of recent research activities within the european project ARISTOTEL," *2011 37th Eur. Rotorcraft Forum, Gallarate, Italy*, pp. 275–293, Sep. 2011.

Pavel, M. D., Malecki, J., DangVu, B., Masarati, P., Gennaretti, M., Jump, M., Smaili, H., Ionita, A., and Zaicek, L., "A retrospective survey of adverse rotorcraft pilot couplings in european perspective," *AHS 2012 American Helicopter Society 68th Annual Forum, Fort Worth (TX)*, pp. 1–14, May 2012.

Perreault, E. J., Kirsch, R. F., and Crago, P. E., "Effects of voluntary force generation on the elastic components of endpoint stiffness," *Experimental Brain Research*, vol. 141, pp. 312–323, 2001, doi:10.1007/s002210100880.

Pintelon, R. and Schoukens, J., *System identification: A frequency domain approach*, IEEE Press, 2001.

Quaranta, G., Masarati, P., and Venrooij, J., "Robust stability analysis: a tool to assess the impact of biodynamic feedthrough on rotorcraft," *AHS 2012 American Helicopter Society 68th Annual Forum, Fort Worth (TX)*, pp. 1–10, May 2012.

Quaranta, G., Masarati, P., and Venrooij, J., "Impact of pilots' biodynamic feedthrough on rotorcraft by robust stability," *J. of Sound and Vibration*, vol. 332, no. 20, pp. 4948–4962, Sep. 2013, doi: 10.1016/j.jsv.2013.04.020.

Raney, D. L., Jackson, E. B., Buttrill, C. S., and Adams, W. M., "The impact of structural vibrations on flying qualities of a supersonic transport," *AIAA Atmospheric Flight Mechanics Conf., Montreal, Canada*, pp. 1–11, Aug. 2001.

Repperger, D. W., "Biodynamic and spasticity reduction in joystick control via force reflection," Tech. Rep. AL/CF-TR-1995-0152, Armstrong Laboratory, Sep. 1995.

Rodchenko, V. V., Zaichik, L. E., and Yashin, Y. P., "Handling qualities criterion for roll control of highly augmented aircraft," *J. of Guidance, Control & Dynamics*, vol. 26, no. 6, pp. 928–934, Nov. 1993.

Schoenberger, R. D. and Wilburn, D. L., "Tracking performance during whole-body vibration with side-mounted and center-mounted control sticks," Tech. Rep. AD0761798, Aerospace Medical Research Lab Wright-Patterson Air Force Base, Apr. 1973.

Schouten, A. C., de Vlugt, E., and van der Helm, F. C. T., "Design of perturbation signals for the estimation of proprioceptive reflexes," *IEEE Trans. on Biomedical Engineering*, vol. 55, no. 5, pp. 1612–1619, May 2008a, doi:10.1109/TBME.2007.912432.

Schouten, A. C., Mugge, W., and van der Helm, F. C. T., "NMClab, a model to assess the contributions of muscle visco-elasticity and afferent feedback to joint dynamics," *J. of Biomechanics*, vol. 41, no. 8, pp. 1659–1667, 2008b.

Schubert, D. W., Pepi, J. S., and Roman, F. E., "Investigation of the vibration isolation of commercial jet transport pilots during turbulent air penetration," Tech. Rep. NASA CR-1560, NASA, Jul. 1970.

Serafini, J., Gennaretti, M., Masarati, P., Quaranta, G., and Dieterich, O., "Aeroelastic and biodynamic modeling for stability analysis of rotorcraft-pilot coupling phenomena," *34th Eur. Rotorcraft Forum, Liverpool, UK*, Sep. 2008.

Sirouspour, M. R. and Salcudean, S. E., "Suppressing operator-induced oscillations in manual control systems with movable bases," *IEEE Trans. on Control Systems Technology*, vol. 11, no. 4, pp. 448–459, 2003, doi:10.1109/TCST.2003.813386.

Snyder, R. G., Ice, J., Duncan, J. C., Hyde, A. S., and Leverett, S. J., "Biomedical research studies in acceleration, impact, weightlessness, vibration, and emergency escape and restraint systems: a comprehensive bibliography," Tech. Rep. CARA Report 63-30, Civil Aeromedical Research Institute, Dec. 1963.

Sövényi, S., *Model-based cancellation of biodynamic feedthrough with a motorized manual control interface*, Ph.D. thesis, University of Michigan, 2005.

Sövényi, S. and Gillespie, R. B., "Cancellation of biodynamic feedthrough in vehicle control tasks," *IEEE Trans. on Control Systems Technology*, vol. 15, no. 6, pp. 1018–1029, Nov. 2007, doi: 10.1109/TCST.2007.899679.

Stroosma, O., van Paassen, M. M., and Mulder, M., "Using the SIMONA research simulator for human-machine interaction research," *AIAA Modeling and Simulation Technologies Conf., Austin (TX)*, pp. 1–8, Aug. 2003.

Tanaka, Y., Yamada, N., Tsuji, T., and Suetomi, T., "Vehicle active steering control system based on human mechanical impedance properties of the arms," *IEEE Trans. on Intelligent Transportation Systems*, vol. 15, no. 4, pp. 1758–1769, Aug. 2014, doi:10.1109/TITS.2014.2312458.

Torle, G., "Tracking performance under random acceleration: effect of control dynamics," *Ergonomics*, vol. 8, no. 4, pp. 481–486, 1965, doi:10.1080/00140136508930829.

van der Helm, F. C. T., Schouten, A. C., de Vlugt, E., and Brouwn, G. G., "Identification of intrinsic and reflexive components of human arm dynamics during postural control," *J. of Neuroscience Methods*, vol. 119, no. 1, pp. 1–14, 2002, doi:10.1016/S0165-0270(02)00147-4.

van der Ouderaa, E., Schoukens, J., and Renneboog, J., "Peak factor minimization using a time-frequency domain swapping algorithm," *IEEE Trans. on Instrumentation and Measurement*, vol. 37, no. 1, pp. 145–147, 1988, doi:10.1109/19.2684.

Velger, M., Grunwald, A. J., and Merhav, S. J., "Suppression of biodynamic disturbances and pilot-induced oscillations by adaptive filtering," *J. of Guidance, Control & Dynamics*, vol. 7, no. 4, pp. 401–409, 1984.

Velger, M., Grunwald, A. J., and Merhav, S. J., "Adaptive filtering of biodynamic stick feedthrough in manipulation tasks on board moving platforms," *J. of Guidance, Control & Dynamics*, vol. 11, no. 2, pp. 153–159, 1988.

Venrooij, J., Abbink, D. A., Mulder, M., van Paassen, M. M., and Mulder, M., "Biodynamic feedthrough is task dependent," *IEEE Int. Conf. on Systems, Man and Cybernetics, Istanbul, Turkey*, pp. 2571–2578, Oct. 2010a, doi:10.1109/ICSMC.2010.5641915.

Venrooij, J., Abbink, D. A., Mulder, M., van Paassen, M. M., and Mulder, M., "A method to measure the relationship between biodynamic feedthrough and neuromuscular admittance," *IEEE Trans. on Systems, Man, and Cybernetics, Part B: Cybernetics*, vol. 41, no. 4, pp. 1158–1169, 2011a, doi:10.1109/TSMCB.2011.2112347.

Venrooij, J., Abbink, D. A., Mulder, M., van Paassen, M. M., Mulder, M., van der Helm, F. C. T., and Bülthoff, H. H., "A biodynamic feedthrough model based on neuromuscular principles," *IEEE Trans. on Cybernetics*, vol. 44, no. 7, pp. 1141–1154, 2014a, doi:10.1109/TCYB.2013.2280028.

Venrooij, J., Mulder, M., Abbink, D. A., van Paassen, M. M., Mulder, M., van der Helm, F. C. T., and Bülthoff, H. H., "Mathematical biodynamic feedthrough model applied to rotorcraft," *IEEE Trans. on Cybernetics*, vol. 44, no. 7, pp. 1025–1038, 2014b, doi: 10.1109/TCYB.2013.2279018.

Venrooij, J., Mulder, M., Abbink, D. A., van Paassen, M. M., van der Helm, F. C. T., Bülthoff, H. H., and Mulder, M., "A new view on biodynamic feedthrough analysis: Unifying the effects on forces and positions," *IEEE Trans. on Cybernetics*, vol. 43, no. 1, pp. 129–142, 2013a, doi:10.1109/TSMCB.2012.2200972.

Venrooij, J., Mulder, M., van Paassen, M. M., and Abbink, D. A., "Relating biodynamic feedthrough to neuromuscular admittance," *IEEE Int. Conf. on Systems, Man and Cybernetics, San Antonio (TX)*, pp. 1668–1673, Oct. 2009, doi:10.1109/ICSMC.2009. 5346935.

Venrooij, J., Mulder, M., van Paassen, M. M., Abbink, D. A., Bülthoff, H. H., and Mulder, M., "Cancelling biodynamic feedthrough requires a subject and task dependent approach," *IEEE Int. Conf. on Systems, Man, and Cybernetics, Anchorage (AK)*, pp. 1670–1675, Oct. 2011b, doi:10.1109/ICSMC.2011.6083911.

Venrooij, J., Mulder, M., van Paassen, M. M., Abbink, D. A., and Mulder, M., "A review of biodynamic feedthrough mitigation techniques," *11th IFAC/IFIP/IFORS/IEA Symposium on Analysis, Design, and Evaluation of Human-Machine Systems, Valenciennes, France*, vol. 11, pp. 316–321, Sep. 2010b, doi:10.3182/20100831-4-FR-2021.00056.

Venrooij, J., Mulder, M., van Paassen, M. M., Abbink, D. A., van der Helm, F. C. T., Mulder, M., and Bülthoff, H. H., "How effective is an armrest in mitigating biodynamic feedthrough?" *IEEE Int. Conf. on Systems, Man, and Cybernetics, Seoul, Korea*, pp. 2150–2155, Oct. 2012, doi:10.1109/ICSMC.2012.6378058.

Venrooij, J., Mulder, M., van Paassen, M. M., Abbink, D. A., van der Helm, F. C. T., Mulder, M., and Bülthoff, H. H., "Admittance-adaptive model-based cancellation of biodynamic feedthrough,"

IEEE Int. Conf. on Systems, Man, and Cybernetics, San Diego (CA), pp. 1946–1951, Oct. 2014c, doi:10.1109/SMC.2014.6974206.

Venrooij, J., Pavel, M. D., Mulder, M., van der Helm, F. C. T., and Bülthoff, H. H., "A practical biodynamic feedthrough model for helicopters," *CEAS Aeronautical Journal*, vol. 4, no. 4, pp. 421–432, Dec. 2013b, doi:10.1007/s13272-013-0083-y.

Venrooij, J., Yilmaz, D., Pavel, M. D., Quaranta, G., Jump, M., and Mulder, M., "Measuring biodynamic feedthrough in helicopters," *37th Eur. Rotorcraft Forum, Gallarate, Italy*, pp. 967–976, Sep. 2011c.

Walden, R. B., "A retrospective survey of pilot-structural coupling instabilities in naval rotorcraft," *American Helicopter Society 63rd Annual Forum, Virginia Beach (VA)*, pp. 897–914, May 2007.

Zaichick, L. E., Desyatnik, P. A., Grinev, K. N., and Yashin, Y. P., "Effect of manipulator type and feel system characteristics on high-frequency biodynamic pilot-aircraft interaction," *28th Int. Congress of the Aeronautical Sciences ICAS, Brisbane, Australia*, pp. 1–7, Sep. 2012.

Appendices

APPENDIX

Fundamentals of biodynamic feedthrough

THIS appendix serves as a point of reference for reviewing the fundamental concepts of biodynamic feedthrough, of which a detailed description is provided in Chapter 3.

Fig. A.1 shows the **biodynamic feedthrough system model**, a conceptual model that shows all elements of a typical biodynamic feedthrough system. Each model block contains a transfer function (indicated with H) describing the dynamics of the system it represents. Tables A.1 and A.2 contain brief descriptions of the elements and signals presented in the BDFT system model.

First, let's consider a human-machine system without the influence of acceleration disturbances: such a system consists of a **human operator (HO)** and a **controlled element (CE)**, e.g., a vehicle. The HO is manually controlling the CE using a **control device (CD)**. The HO generates control commands by comparing the current state of the CE $y_{cur}(t)$ with a goal state $y_{goal}(t)$. Based on differences $y_{err}(t)$ between these two states the HO's **central nervous system (CNS)** – which is responsible for all cognitive control commands – formulates a voluntary control command, described here as a cognitive supra-spinal input $n_{cog}(t)$, which is transmitted neurally to the **neuromuscular system (NMS)**. The NMS represents the dynamics of the limb connected to the CD and contains body parts such as

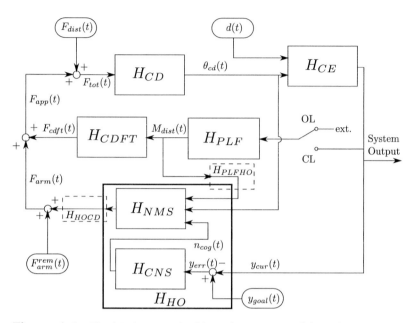

Figure A.1: The biodynamic feedthrough system model. A human operator (HO) controls a controlled element (CE) using a control device (CD). Motion disturbances $M_{dist}(t)$ are coming from the platform (PLF). The feedthrough of $M_{dist}(t)$ to involuntary applied forces $F_{arm}(t)$ and involuntary control device deflections $\theta_{cd}(t)$ is called biodynamic feedthrough (BDFT). The feedthrough of $M_{dist}(t)$ to inertia forces $F_{cdft}(t)$ is called control device feedthrough (CDFT). $F_{app}(t)$ is the sum of the forces applied to the control device by the HO. The HO consists of a central nervous system (CNS) and a neuromuscular system (NMS). The connection between the HO and the environment is governed by two 'interfaces', H_{PLFHO} and H_{HOCD}. The CE and PLF can form an open-loop (OL) or closed-loop (CL) system.

bones, muscles, etc. The CNS includes the corticospinal tract or 'upper motor neurons', the NMS includes the spinal tract or 'lower motor neurons'. The NMS, which in this thesis is assumed to be a human arm, exerts a force $F_{arm}(t)$ on the CD. The control device deflections $\theta_{cd}(t)$ form the control input for the CE. Note that the CE can also be perturbed by a disturbance signal $d(t)$, for which the HO should compensate.

Table A.1: BDFT system model elements.

Element	Description
H_{CD}	Control device dynamics
H_{CDFT}	Control device feedthrough dynamics; effect of M_{dist} on CD
H_{CE}	Controlled element dynamics; system under control by HO
H_{CNS}	Central nervous system dynamics; brain and spinal cord of HO
H_{HO}	Human operator dynamics
H_{HOCD}	Interface dynamics between HO and CD; e.g., grip dynamics, armrest
H_{NMS}	Neuromuscular system dynamics; muscles, bones, etc., of HO
H_{PLF}	Platform dynamics; source of motion disturbance M_{dist}
H_{PLFHO}	Interface dynamics between PLF and HO; e.g., seat dynamics

Table A.2: BDFT system model signals.

Signal	Description
$\theta_{cd}(t)$	Control device deflection (position)
$d(t)$	External disturbance on CE
$F_{app}(t)$	Force applied on control device (externally)
$F_{arm}(t)$	Force applied by the human operator (here, through arm)
$F_{arm}^{rem}(t)$	Force applied as result of operator remnant
$F_{cdft}(t)$	Control device feedthrough force
$F_{dist}(t)$	Force disturbance on control device
$F_{tot}(t)$	Total force on control device
$M_{dist}(t)$	Motion disturbance, originating from PLF
$n_{cog}(t)$	Cognitive (voluntary) control signal (neural commands)
$y_{cur}(t)$	Current state of CE
$y_{err}(t)$	Difference between goal and current state of CE
$y_{goal}(t)$	Goal state of CE

The representation of the manually controlled human-machine system can be extended to account for the effect of accelerations. These typically originate from the motion of a vehicle, referred to as the **platform (PLF)**. The acceleration signal, coming from H_{PLF}, is called the motion disturbance signal $M_{dist}(t)$. The influence of the motion disturbance signal on the human-machine system is modeled through two effects: first, the mass of the control device m_{cd} converts the PLF accelerations into inertial forces (also known as fictitious forces or d'Alembert forces) $F_{cdft}(t)$. This effect is described in the H_{CDFT} block and is labeled **control device feedthrough (CDFT)**. Secondly, the PLF accelerations are transferred through the body of the HO and induce unintentional motions in the limb that is in contact with the CD, thereby leading to unintentional forces applied to the control device and – if the control device is movable, i.e., not rigid – these result in involuntary deflections of the control device. The generation of both involuntary forces and involuntary deflections is what is defined here as **biodynamic feedthrough (BDFT)**:

> **Biodynamic feedthrough**
>
> The transfer of accelerations through the human body during the execution of a manual control task, causing involuntary forces being applied to the control device, which may result in involuntary control device deflections.

For the control device feedthrough (CDFT) the following definition is proposed:

> **Control device feedthrough**
>
> The transfer of accelerations through the control device mass, resulting in inertial forces being applied to the control device.

When the operator is both on board *and* controlling the vehicle, a connection exists between the CE and the PLF, as the HO's inputs to the CE affect the PLF's motion. This situation is referred to as a **closed-loop (CL)** BDFT system, a type of BDFT system that can lead to weakly damped or unstable oscillations. The alternative situation, an **open-loop (OL)** BDFT system, occurs if the HO is a passenger on board of a moving vehicle and engaged in a manual control task other than control of that same vehicle. Both closed-loop and open-loop BDFT systems are important and practically relevant. In the model these two types are included through a switch which can either open or close the loop between CE and PLF. In the open-loop case the PLF receives inputs from outside the human-machine system considered here. These external inputs are indicated in Fig. A.1 as 'ext.'

The two dashed boxes shown in Fig. A.1 are the two 'interfaces' that govern the connection between the operator and the environment. These blocks are indicated with dashed lines as they do not contain the dynamics of a single physical element, but dynamics that are influenced by several other systems. In the analysis, the interface dynamics are therefore often lumped with other dynamics. The **PLFHO interface** describes the dynamics of the connection between the PLF and the HO, influenced by, for example, seat suspension or seat belts, representing the interaction dynamics with the body of the HO. These dynamics are sometimes referred to as the 'seat transmissibility' and they determine how accelerations enter the operator's body. The **HOCD interface** describes the dynamics of the connection between HO and CD, e.g., grip visco-elasticity or the effect of an armrest. This interface determines how limb motions result in – voluntary and involuntary – forces $F_{arm}(t)$.

Finally, Fig. A.1 shows the addition of a force disturbance signal, $F_{dist}(t)$. This force signal is applied to the control device and is used to obtain an estimate of the dynamics of the human limb in contact with the CD. One of the goals of this thesis is to investigate the influence of these dynamics on BDFT. To describe the adaptive dynamics of human limbs the neuromuscular admittance is used. Admittance can be defined as [Abbink et al., 2011]:

Neuromuscular admittance

The causal dynamic relationship between the force acting on the limb (input) and the position of the limb (output).

The admittance, when determined by linear time-invariant (LTI) estimation techniques, shows properties of a mass-spring-damper system due to visco-elastic properties of the muscle and the limb inertia, as well as higher-order dynamics due to reflexive activity and grip dynamics. Some of the physical properties underlying the admittance can be assumed to be time-invariant, such as inertia, reflexive time delays, but others are highly adaptive such as the reflexive activity and muscle co-contraction.

A large admittance means that a force acting on a limb results in large position deviations, which would occur for compliant limbs; a small admittance means a force results in small position deviations, which would occur for stiff limbs.

Neuromuscular admittance can be structurally varied using the following three control tasks (also referred to as the classical tasks):

- The position task (PT) (or stiff task), in which the instruction was to keep the position of the control device in the centered position, that is, to "resist the force perturbations as much as possible".

- The force task (FT) (or compliant task), in which the instruction was to minimize the force applied to the control device, that is, to "yield to the force perturbations as much as possible".

- The relax task (RT), in which the instruction was to relax the arm while holding the control device, that is, to "passively yield to all perturbations".

The human operator needs to set his/her neuromuscular properties differently for optimal control of each of these control tasks. The PT is a task for which the best performance is achieved by being very stiff (i.e., a small admittance), the FT requires the operator to be very compliant (i.e., a large admittance). The RT is intended to yield an admittance which gives an indication of the passive dynamics of the neuromuscular system.

Practical guidelines for biodynamic feedthrough mitigation

Several ways of mitigating biodynamic feedthrough were discussed in this thesis. Mostly, however, the discussion was limited to theoretical aspects. This appendix aims to provide several practical guidelines on how one could approach the mitigation of biodynamic feedthrough in actual vehicles. The approach proposed here is divided in four steps:

- Step 1: Identifying the BDFT problem

- Step 2: Indexing the possible mitigation approaches

- Step 3: Performing the approach trade-off

- Step 4: Implementing the selected mitigation approach

In the following, each step will be elaborated.

B.1 Step 1: Identifying the BDFT problem

Before making any attempt at BDFT mitigation, the actual problem needs to be properly identified. The goal of this step is to gain as much knowledge as possible regarding the characteristics of the BDFT problem one is dealing with.

In many practical cases, a BDFT problem will reveal itself through the experience of operators that are suffering from involuntary control inputs. It is important that operators are made aware of the existence of the BDFT phenomenon, are trained to identify BDFT events and are encouraged to report those events. This is not a trivial aspect, given that the topic of BDFT is not commonly treated in the educational curriculum of many professional operators.

There are also situations where practical experience with BDFT is absent or unavailable, for example, when investigating the *possible* occurrence of BDFT in a vehicle's design phase. In that case, an experimental campaign may be the only possible route to obtain indications whether BDFT may turn out to be a relevant problem for the vehicle in question. BDFT models, describing BDFT dynamics in a similar environment as the one that is under consideration, may be very helpful in identifying possible BDFT proneness.

Amongst the information that would be required to properly deal with a BDFT problem is knowledge on the condition in which BDFT occurs (at which vehicle speed, in which loading conditions, etc.). Furthermore, it is very helpful to know at which frequencies the involuntary inputs manifest themselves. Measurements on human subjects in a representative simulated environment or in the actual vehicle are highly useful and increase the chances of success in the following steps.

B.2 Step 2: Indexing the possible mitigation approaches

Once the characteristics of the BDFT problem are identified, it is important to choose the appropriate strategy to mitigate it. There are several BDFT mitigation approaches available, Chapter 6 provides a detailed discussion of each approach, including benefits and disadvantages.

The goal of this step is to obtain a list of all possible approaches, discarding those that are not feasible or are undesired. For example, in some cases, the armrest may be discarded as a feasible solution, as there may be no space available at the operator's station to install one (see Section 6.2.4). Or regulations may stipulate that no adaptations can be made to the control device dynamics, ruling out that option of BDFT mitigation (Section 6.2.5). An option that is likely to appear on the list of possible mitigation approaches is 'neuromuscular adaptation' (Section 6.2.3), as this approach requires no adaptation to the vehicle at all. A method that is less likely to survive the selection is 'minimize platform acceleration' (Section 6.2.1), as this may be impossible to achieve, or only at great cost and with severe repercussion in other aspects of the vehicle's performance.

In some cases, for example during the design phase of a vehicle, the list of possible mitigation approaches may be long, as many adaptations to the vehicle are still possible. In other cases, for example when mitigating BDFT in a vehicle that is already in operation, the amount of possible mitigation approaches may be small, possibly limited to one or two.

B.3 Step 3: Performing the approach trade-off

In this step the results of the two previous steps will be combined. Putting together the knowledge on the BDFT problem with the possible ways of dealing with the problem allows us to make a proper choice. The reason for separating the listing of possible options from selecting the desired option is that it reduces the chances of overlooking viable options. For example, the occurrence of BDFT in rotorcraft is often mitigated using either notch filters, adaptations of the control device and through procedural mitigations [Walden, 2007]. As was discussed in Section 3.10.3, the option of adapting the interface dynamics between the rotorcraft and the body of the pilot (e.g., by changing seat dynamics) is rarely considered, even though this may prove to be a perfectly valid approach that is worth considering. In order to prevent valid options from not being exploited it is important to first obtain a list of what is *possible* (step 2) before deciding what is *best* (step 3).

In the likely case that the selection of possible approaches yielded two or more options, a trade-off needs to be made. Factors in this trade-off will be the available budget, available knowledge, previous experiences, etc. This is also where the obtained knowledge on the BDFT problem comes into play. If the BDFT problem is unlikely to occur, not directly critical, or easily handled by the human operator, a simple and cost-effective measures such as 'neuromuscular adaptation' may be preferred. In other situations, where the occurrence of BDFT is likely and highly critical (e.g., surgery on board of moving vehicles), a method to actively remove BDFT effects may be required, such as 'model-based BDFT cancellation'.

One of the findings of this thesis that is of particular importance when considering different options for BDFT mitigation is that biodynamic feedthrough is a variable relationship, varying both between different persons (between-subject variability), as well as within one person over time (within-subject variability). Hence, what works in one situation, may fail in another. A thorough evaluation of the method is therefore recommended (see step 4).

The following will provide some general consideration regarding each of the possible solution types. This serves as an addition to the discussion of their benefits and disadvantages that was provided in Chapter 6.

Minimizing platform accelerations   When considering this approach, it may be very useful to obtain an open-loop BDFT model (see, for example, Chapter 4, Chapter 5 and [Venrooij et al., 2013b]). The model should provide a representative description of BDFT dynamics, preferably for a range of settings of the neuromuscular system, for the operator station that is under consideration. This model can than be combined with a vehicle dynamics model, of which the dynamics can be adapted. The effectiveness of these changes can then be accurately assessed, for example by performing a robust stability analysis [Quaranta et al., 2013].

PLF-HO interface design   This approach was not experimentally addressed in this thesis. More research is required to explore the influence of, e.g., seat dynamics on BDFT. Interesting work on BDFT mitigation by adapting the PLF-HO interface was done by Schubert et al. [1970], Dimasi et al. [1972] and more recently by Humphreys et al. [2014]. Also, the work on the influence of seat dynamics on whole-body vibrations (e.g., [Griffin, 1978]) may be relevant when delving into this topic.

Neuromuscular adaptation   If the range of frequencies where BDFT problems manifest themselves is known, the BDFT transfer dynamics, such as shown in Fig. 2.8 on page 54, may be helpful in determining which setting yields the largest benefit in minimizing BDFT. As a rule of thumb, stiff behavior is beneficial at low frequencies, relaxed or compliant behavior is beneficial at high frequencies. The specifics of the optimal setting strongly depend on control device dynamics, disturbance direction, etc.
Neuromuscular adaptation is a powerful BDFT mitigation method. In addition, it seems to come 'for free', as it requires no adaptation to the vehicle. However, an important disadvantage of this method is that it relies on the operator to mitigate a complex involuntary phenomenon while being engaged in a manual control task. For example, the instruction to 'reduce grip' to mitigate BDFT events [Walden, 2007] might fail in stressful or critical situations, exactly when it is most needed. In fact, any procedural instruction provided to the operator may be counterintuitive or undesired when the need arises to apply it. This may leave such procedural instructions ignored and the BDFT problem not handled.

HO-CD interface design   The work presented in this thesis has shown that an armrest is both an effective and efficient tool in mitigating BDFT, and this is therefore an option that is worth considering. Two main requirements need to be satisfied. The firs one is that an armrest actually provides support when controlling the control device in question (this is less likely for, e.g., a steering wheel, and more likely for a side-stick). The second is that there is enough space at the operator's station to install such a device. More research would be required to understand the exact influence of the armrest's location and properties on the BDFT mitigation effectiveness. Chapter 7 may provide inspiration on how to perform such measurements and analyze the results.

Control device design   If data on both the BDFT dynamics and neuromuscular admittance (or force disturbance feedthrough dynamics) are available, the influence of changing the control device dynamics can be calculated, using the approach detailed in Section 3.9. Also a physical BDFT model can be used to study the influence of variations in the control device dynamics (see Section 4.8.4). It should be kept in mind, however, that adapting the control device dynamics may have an influence on other aspects of controllability. Also, as was shown in Section 3.8.3, the adaptation of the control device dynamics may have a negligible influence on the BDFT dynamics if the neuromuscular admittance of the operator is much stiffer than the control device dynamics (which may occur when the operator is stressed).

Signal filtering   This approach was not experimentally addressed in this thesis. If the BDFT problem is confined to a very narrow range of frequencies, a notch filter may be applied. It should be kept in mind that adding filters may also introduce additional delay in the control loop, which may lead to other control problems. More research would be required to investigate how novel filtering techniques may be used in BDFT mitigation. The work by Velger et al. [1984, 1988] may be a convenient starting point.

Model-based cancellation   For this approach, there are two main options: force cancellation and signal cancellation. If one decides to go for the former, one would need a B2FOL dynamics model. An advantage of such a model is that it is independent of control device dynamics. An important disadvantage of force cancellation is that it requires an active stick. If such a stick is not already present in the vehicle, the additional cost, weight and complexity of adding one may make it worth considering signal cancellation instead. Signal cancellation requires a B2P dynamics model. Such a model is dependent on the control device dynamics. The model can be based on a B2FCL model (as was done in Chapter 5). The most important disadvantage of this approach is that the human is not directly informed about the controller's activity, which may prevent or delay the operator noticing errors in the corrective control due to, e.g., inaccuracies in the BDFT model.

In both cases of model-based cancellation, one should decide on the level of generality of the BDFT model. Individual models have shown to be superior to more general models (see, for example, Section 5.5 and Section 6.5.5). However, in some cases the added complexity of obtaining such models may make the choice for a simpler, more general model preferable.

B.4   Step 4: Implementing the selected mitigation approach

The final step is the implementation of whichever method was selected. Often, such an implementation will require simulations or experiments to tune the details of the approach and validate its effectiveness. Chapter 6, Chapter 7 and Chapter 8 provide examples of how an approach can be evaluated.

After the mitigation approach is in place and operational, it is important to evaluate its effectiveness in practice to make sure the BDFT problem is actually solved. Also, it is important to ensure that the alterations made do not influence other aspects of control, such as the voluntary control inputs. Finally, it is recommended to make the results of such an evaluation available to others. By sharing the successes and failures of BDFT mitigation in practice we can improve the effectiveness and efficiency of future BDFT mitigation methods.

Samenvatting

Meten, modelleren en mitigeren
van biodynamische doorvoer

Joost Venrooij

VOERTUIGVERSNELLINGEN hebben verschillende uitwerkingen op het menselijk lichaam. In sommige gevallen kunnen ze leiden tot onvrijwillige bewegingen van ledematen zoals armen en handen. Als iemand op hetzelfde moment bezig is met het uitvoeren van een handmatige stuurtaak kunnen deze onvrijwillige bewegingen van de ledematen leiden tot onvrijwillige stuurkrachten en stuurinvoer. Dit fenomeen heet biodynamische doorvoer (BDFT[a]). Het is bekend dat BDFT kan optreden bij het besturen van verschillende soorten voertuigen, zoals helikopters, vliegtuigen, elektrische rolstoelen en hydraulische graafmachines.

Het feit dat BDFT comfort vermindert, stuurprestaties verslechtert en de veiligheid in gevaar kan brengen voor diverse voertuigen en onder diverse omstandigheden heeft ertoe geleid dat al vele onderzoeken zijn uitgevoerd om BDFT effecten te meten, modelleren en mitigeren. Ondanks de aandacht die aan BDFT besteed is over

---

[a]In deze samenvatting wordt voor afkortingen de Engelse schrijfwijze gehanteerd. Zie de 'Nomenclature' op pagina xv voor de Engelse betekenis.

de afgelopen decennia, blijven vele vragen omtrent BDFT onbeant-woord. Het is met name duidelijk geworden dat BDFT een complex fenomeen is, waarin verschillende factoren een rol spelen. Boven-dien begrijpen we de invloed van vele van deze factoren nog maar nauwelijks. Het is bekend dat BDFT dynamica afhangt van voer-tuigdynamica en stuurorgaandynamica, maar ook van factoren zo-als stoeldynamica, verstoringsrichting, verstoringsfrequentie en de aanwezigheid van stoelgordels en/of armsteunen.

De meest complexe en invloedrijkste factor in BDFT is het menselijk lichaam: de menselijke lichaamsdynamica bepaalt in hoge mate in hoeverre de voertuigacceleraties tot onvrijwillige bewegingen van ledematen leiden. De menselijk lichaamsdynamica verschilt tus-sen personen met verschillende lichaamsgrootte en -gewicht, maar varieert ook gedurende de tijd voor één persoon. Het is bekend dat mensen het neuromusculaire systeem van hun lichaam aanpas-sen door spiercocontractie en het moduleren van reflexieve activi-teit als reactie op, onder andere, taakinstructie, werkdruk en ver-moeidheid. Dit maakt BDFT tot een variabele dynamische rela-tie, die niet alleen varieert tussen verschillende personen (tussen-persoonsvariabiliteit) maar ook gedurende de tijd voor één enkel persoon (binnen-persoonsvariabiliteit).

Het onderzoeksdoel van het werk dat in dit proefschrift wordt ge-presenteerd was **het vergroten van onze kennis van BDFT ten-einde een effectieve en efficiënte methode te ontwikkelen voor het mitigeren van BDFT problemen**. Dit proefschrift behandelt verschillende aspecten van biodynamische doorvoer, maar het werk concentreert zich op **de invloed van de variabele neuromuscu-laire dynamica op BDFT dynamica**. De aanpak van het onderzoek bestaat uit drie delen: ten eerste is een methode ontwikkeld om BDFT op nauwkeurige wijze te meten. Vervolgens zijn verschil-lende BDFT modellen ontwikkeld, gebaseerd op verscheidene prin-cipes, die het BDFT fenomeen beschrijven. Tot slot, gebruikmakend van de inzichten die in de voorgaande stappen zijn verworven, is een nieuwe aanpak voor het mitigeren van BDFT ontwikkeld.

Ten einde een goed begrip te ontwikkelen betreffende de invloed van neuromusculaire dynamica op BDFT dynamica moeten beide gelijktijdig gemeten worden. Een goede maat voor het beschrijven van neuromusculaire dynamica is de neuromusculaire admittantie. Admittantie is een dynamische eigenschap van een ledemaat, die gekarakteriseerd wordt door de relatie tussen kracht-invoer en positie-uitvoer van een ledemaat. Dit proefschrift beschrijft een methode die het mogelijk maakt BDFT en neuromusculaire admittantie gelijktijdig te meten. Met behulp van deze methode zijn verschillende inzichten verworven betreffende de relatie tussen deze twee soorten dynamica. Dit proefschrift beschrijft in detail de experimenten en de resultaten waarmee de methode gevalideerd is.

In de experimenten werd de admittantie gevarieerd door middel van drie verschillende stuurtaken: de positie taak (PT), of 'stijve taak', met de taakinstructie de positie-uitwijkingen van het stuurorgaan te minimaliseren; de kracht taak (FT), of 'compliante taak', met de taakinstructie de kracht die wordt uitgeoefend op het stuurorgaan te minimaliseren; en de relax taak (RT), met de instructie de arm te ontspannen terwijl het stuurorgaan wordt vastgehouden. Door het volgen van de PT, FT, en RT taakinstructies brachten de proefpersonen verschillende instellingen van hun neuromusculaire systeem tot stand: respectievelijk een maximaal stijve instelling (een lage admittantie), een maximale compliante instelling (een hoge admittantie) en een passieve instelling.

De resultaten van de experimentele validatie van de methode laten zien dat de methode succesvol was in het gelijktijdig meten van admittantie en BDFT. Gebaseerd op de gemeten variatie in BDFT dynamica en neuromusculaire admittantie kan geconcludeerd worden dat **de BDFT dynamica sterk afhankelijk is van de neuromusculaire admittantie**.

In de literatuur is er weinig overeenstemming betreffende de definities, nomenclatuur en wiskundige beschrijvingen die gebruikt moeten worden in het onderzoek naar BDFT problemen. In dit proefschrift is een raamwerk ontwikkeld voor BDFT analyse dat als basis kan dienen voor het bestuderen, bediscussiëren en begrijpen van BDFT en gerelateerde problemen. Gebruikmakend van het raamwerk kunnen oude en nieuwe BDFT onderzoeken worden ge(her)interpreteerd, geëvalueerd en vergeleken. Tevens van belang is dat het raamwerk zelf ook kan worden gebruikt om nieuwe inzichten te vergaren met betrekking tot het BDFT fenomeen.

Binnen het raamwerk wordt onderscheid gemaakt tussen het genereren van onvrijwillige stuurkrachten en het genereren van onvrijwillige stuurbewegingen (posities). In dit proefschrift worden deze twee effecten respectievelijk BDFT tot krachten (B2F) en BDFT tot posities (B2P) genoemd. B2P is het onderwerp van het merendeel van de bestaande BDFT literatuur. De introductie van de B2F dynamica leidt tot enkele waardevolle nieuwe inzichten. De B2F dynamica kan op twee manieren worden gedefinieerd: BDFT tot krachten in open lus (B2FOL) en BDFT tot krachten in gesloten lus (B2FCL). Beide vormen van B2F dynamica beschrijven verschillende aspecten van het BDFT fenomeen.

Het raamwerk bevat ook wiskundige relaties die beschrijven hoe verschillende soorten dynamica zich tot elkaar verhouden. Deze relaties zijn gevalideerd met behulp van experimentele gegevens. De conclusie van deze validatie is dat **het nut van het raamwerk is bewezen, zowel bij het interpreteren van BDFT studies uit het verleden, als bij het vergaren van nieuwe inzichten.**

De huidige BDFT modellen kunnen worden ingedeeld in grofweg twee groepen: fysische BDFT model en zwarte-doos BDFT modellen. Beide hebben als doel BDFT dynamica te beschrijven, maar doen dat op een verschillende manier. De fysische modellen zijn erop gericht een fysische representatie te verschaffen van het BDFT fenomeen, daarbij gebruikmakend van a-priori kennis en fysische principes. Zwarte-doos modellen zijn erop gericht een efficiënte BDFT beschrijving te verschaffen op 'eindpunt niveau'.

In dit proefschrift zijn twee nieuwe BDFT modellen ontwikkeld. De eerste is een fysisch model gebaseerd op neuromusculaire principes. Dit model heeft voornamelijk als doel het inzicht in de relatie tussen admittantie en BDFT te vergroten. Het tweede model is een mathematisch model, welke als doel heeft het gat te dichten tussen de traditionele fysische en zwarte-doos modellen.

Een validatie van het fysische model laat zien dat **het fysische model een nauwkeurige fysische beschrijving geeft van de BDFT dynamica, die onze kennis van het BDFT fenomeen vergroot.** Een van de belangrijkste bijdragen van dit model is het vermogen om zowel tussen-persoons- als binnen-persoonsvariabiliteit in BDFT te beschrijven, een aspect dat in vele bestaande BDFT modellen ontbreekt.

Het tweede model, het mathematische model, is ontworpen met behulp van asymptootmodellering, wat een structurele methode verschaft om de overdrachtsfunctie van een model te genereren. Het resultaat is een zeer accuraat BDFT model met geringe complexiteit, wat een betrouwbare parameterschatting en een eenvoudige implementatie mogelijk maakt. Een studie naar de modelprestaties leidde tot de conclusie dat **het mathematische BDFT model zeer accuraat is, verschillende zwarte-doos modellen overtreft en eenvoudiger in het gebruik is dan fysische modellen.** Bovendien kan worden geconcludeerd dat **asymptootmodellering een succesvolle aanpak is in het ontwerpen van een accurate en veelzijdige modelstructuur voor het mathematische BDFT model.** De methode is hoogstwaarschijnlijk ook bruikbaar in andere modelleringsproblemen.

Met behulp van het BDFT systeem model zijn de beschikbare BDFT mitigeringsmethoden in kaart gebracht en geëvalueerd. In totaal zijn zeven verschillende mitigeringstypen, elk met één of meerdere mitigeringsaanpakken, geïdentificeerd en besproken. Twee mitigeringstypen worden beschouwd als meest veelbelovend. Maatregelen van het eerste mitigeringstype – passieve ondersteunings- en bedwangsystemen (zoals armsteunen en stoelgordels) – worden reeds toegepast. Onderzoek heeft uitgewezen dat deze maatregelen niet afdoende zijn om BDFT volledig te mitigeren. Het tweede mitigeringstype is model-gebaseerde BDFT annulering, waar gebruik wordt gemaakt van een BDFT model om een annuleringssignaal te berekenen. Deze aanpak heeft al wat aandacht ontvangen in de literatuur, maar slechts enkele experimentele implementaties zijn beschreven. Gebruikmakend van een methode genaamd optimale signaalannulering werd bewezen dat **signaalannulering een veelbelovende mitigeringsmethode voor BDFT problemen is als het model aangepast kan worden aan zowel de persoon als de taak.** Aanpassing aan de taak, of beter gezegd, aanpassing aan de neuromusculaire dynamica van de persoon, is van specifiek belang.

De effectiviteit van een armsteun, een voorbeeld van een passief ondersteuningssysteem, in het mitigeren van BDFT effecten is ook experimenteel onderzocht. Het resultaat laat zien dat, in het algemeen, de aanwezigheid van een armsteun de BDFT problemen vermindert. Bovendien leiden de resultaten tot het nieuwe inzicht dat het effect van de armsteun sterk varieert met verstoringsfrequentie en neuromusculaire admittantie. De belangrijkste conclusie van de analyse is dat **een armsteun een effectief middel is in het mitigeren van biodynamische doorvoer**. De installatie van een armsteun zal in veel stuurtaken waar BDFT nu een probleem vormt voldoende zijn om een adequate taakprestatie te bereiken en gesloten lus oscillaties te voorkomen. Dit maakt de armsteun een waardevol alternatief voor complexere mitigeringsmethoden.

Tot slot is een nieuwe aanpak voor BDFT mitigatie ontwikkeld: admittantie-adaptieve model-gebaseerde signaalannulering. Wat deze methode van andere bestaande BDFT mitigeringsmethoden onderscheidt is dat deze methode rekening houdt met de aanpassing van de neuromusculaire dynamica van het menselijk lichaam. De aanpak is conceptueel getest in een experimentele opstelling waar proefpersonen in een bewegingssimulator met een gesimuleerd voertuig door een virtuele tunnel vlogen. Door de stuurprestaties te beoordelen met en zonder bewegingsverstoringen en met en zonder de actieve signaalannulering zijn de prestaties van de aanpak in kaart gebracht. De resultaten laten zien dat **de admittantie-adaptieve model-gebaseerde signaalannulering methode succesvol was en grotendeels de negatieve effecten van BDFT op de stuurprestaties een stuurinspanningen heeft verwijderd**.

Een synthese van de resultaten die in dit proefschrift zijn gepresenteerd leidt tot volgende conclusies:

- De huidige BDFT onderzoeksomgeving is gefragmenteerd en BDFT problemen worden vaak onderzocht van geval tot geval. Meer overeenstemming in de definities, nomenclatuur en wiskundige beschrijvingen zou de kennis van biodynamische doorvoer ten goede komen, de communicatie tussen onderzoekers verbeteren en het vergelijken van studies vergemakkelijken.

- De neuromusculaire dynamica, en met name de variatie daarin, heeft een belangrijke invloed op de biodynamische doorvoer waarmee rekening gehouden dient te worden bij het meten, modelleren en mitigeren van biodynamische doorvoer effecten.

- Er zijn vele mogelijke manieren om biodynamische doorvoer te modelleren. Welke daarvan de voorkeur heeft is sterk afhankelijk van het beoogde gebruik van het model. In het algemeen zouden biodynamische doorvoer modellen zodanig ontworpen moeten worden dat specialisten nieuwe inzichten op gedetailleerde wijze in het model kunnen opnemen, terwijl het resultaat voldoende praktisch en bruikbaar blijft voor gebruik door een bredere gebruikersgroep.

- Het mitigeren van biodynamische doorvoer door middel van model-gebaseerde signaalannulering is een krachtige en veelzijdige aanpak, maar zal alleen succesvol zijn als het biodynamische doorvoer model aangepast wordt aan zowel de bestuurder als de stuurtaak. Meer onderzoek is nodig om de obstakels te overwinnen die momenteel de toepassing van model-gebaseerde signaalannulering in voertuigen belemmeren.

Tot slot, voor toekomstig onderzoek worden de volgende aanbevelingen gedaan:

- Nieuwe identificatiemethoden moeten worden ontwikkeld om het meten van BDFT dynamica in meer natuurlijke stuurtaken mogelijk te maken.

- Het verbreiden van het gebruik van het raamwerk voor BDFT analyse, zoals dat is voorgesteld in dit proefschrift, zou moeten worden aangemoedigd om het raamwerk te verbeteren en uit te breiden.

- Toekomstig onderzoek naar BDFT modellering moet worden toegespitst op het ontwerpen van modellen met beperkte complexiteit, zonder op kwaliteit in te boeten.

- Praktisch onderzoek naar het optreden van BDFT in werkelijke voertuigen is noodzakelijk om effectieve oplossingen te vinden voor de BDFT problemen om ons heen.

Acknowledgments

I firmly believe that thanking someone personally is a much greater sign of gratitude than mentioning a person's name in the acknowledgment section of a book. However, this thesis would not be complete without such a section, as I do owe gratitude to a large number of people for supporting me during the work that led up to this thesis.

First of all, I would like to thank my promotors Max Mulder and Heinrich Bülthoff for giving me the opportunity and support to pursue my Ph.D. degree in a collaborative project between Delft University of Technology and Max Planck Institute for Biological Cybernetics. Max, you have played a pivotal role in my research endeavors ever since I walked into your office looking for a topic for my Master's thesis. Your approachability, people skills and loose rein supervision style has made working with you motivating, productive and enjoyable. Heinrich, I would like to thank you for supporting my research and, at the same time, allowing me to gain professional experiences that I would otherwise not have been able to obtain.

I owe great gratitude to David Abbink, both for his invaluable scientific contribution to the research and for being such a remarkably energetic and warmhearted person. I would also like to thank Mark Mulder, for being a tireless reader of my drafts and his calm and pragmatic approach to all sorts of problems. I further extend my gratitude to René van Paassen, for his valuable feedback and for wielding his renowned all-round analytical abilities towards improving my work. I am thankful for the supervision I received from Frans van der Helm, whose expert advice has played an instrumental role in setting out the research path leading up to this thesis.

I would like to thank Olaf Stroosma and Alwin Damman for providing the necessary technical support during the experimental work. I am also indebted to all the participants that took part in the experiments. Thank you for putting up with being shaken around for hours. Needless to say that without you this thesis would not have existed. I hope that by writing this thesis I have partially made up to my promise that with your dedication you have made a contribution to science.

I gratefully acknowledge Frank Nieuwenhuizen and Joost Ellerbroek for kindly providing the style templates that this thesis is based on.

I would like to thank my colleagues at the Control and Simulation department and the Haptics lab at Delft University of Technology. A special word of thanks goes out to Marilena Pavel, for sharing an office with me and inviting me to contribute to the ARISTOTEL project.

At Max Planck Institute for Biological Cybernetics, I would like to thank my colleagues, in particular the myCopter group, for providing useful feedback on my work, and the Motion Perception and Simulation group, for allowing me to dedicate much more time than anticipated on the writing of my thesis. Special thanks go out to Paolo Pretto and Frank Nieuwenhuizen, for being great officemates and a source of advice on the ways of the institute.

One person that does not fit in any of the categories covered above, but that I do need to mention here is Erwin Boer. He has played an influential role in encouraging me to pursue a Ph.D degree, which has been a decision that I have not regretted and most likely never will.

Finally, I would like to thank my friends and family, especially my parents, my brother Ward and his wife Diana for their continuous support. And, most importantly, I would like to express my sincere gratitude to Rocío, who has been the closest witness of both the highs and the lows of my Ph.D. project. Thank you for being there every step of the way and providing me with your unwavering loving support and encouragement.

Joost Venrooij   *Tübingen, February 2014*

Curriculum Vitae

Joost Venrooij was born on November 1st, 1983, in Vught, the Netherlands. From 1996 to 2002 he attended Maurick College in Vught, where he obtained his Gymnasium diploma (cum laude).

In 2002 he enrolled at the Faculty of Aerospace Engineering at Delft University of Technology (Delft, the Netherlands). In academic year 2005/06 he was board member of the Society for Aerospace Engineering Students VSV 'Leonardo da Vinci'. As part of his M.Sc. studies he was an intern at LUEBEC (La Jolla, CA, USA) for three months, followed by a second internship at Eurocopter (Ottobrunn, Germany) for five months. In August 2009 he obtained his M.Sc. degree (cum laude). In addition, he received an accreditation for completing Delft University of Technology's Honours Programme. In his M.Sc. thesis he investigated the relationship between neuromuscular admittance and biodynamic feedthrough.

After completing his M.Sc. studies he traveled and then worked as research employee at Entropy Control Inc. (La Jolla, CA, USA) for three months. In January 2010 he started his Ph.D. project, in which he continued his research into biodynamic feedthrough. This

Ph.D. project was a collaboration between the Faculty of Aerospace Engineering and the Faculty of Mechanical Engineering of Delft University of Technology and Max Planck Institute for Biological Cybernetics (Tübingen, Germany). He completed his first year of his Ph.D. research at Delft University, the remaining time at the Max Planck Institute.

Currently, Joost is project leader of the Motion Perception and Simulation research group at Max Planck Institute for Biological Cybernetics.

Publications

In the following, only the publications related to biodynamic feed-through are listed.

Journal publications

Venrooij, J, Mulder, M, Mulder, M, Abbink, DA, van Paassen, MM, van der Helm, FCT & Bülthoff, HH (–), "Admittance-adaptive Model-based Approach to Mitigate Biodynamic Feedthrough", submitted to *IEEE Transactions on Cybernetics*.

Venrooij, J, van Paassen, MM, Mulder, M, Abbink, DA, Mulder, M, van der Helm, FCT & Bülthoff, HH (September 2014), "A Framework for Biodynamic Feedthrough Analysis – Part II: Validation and Application", *IEEE Transactions on Cybernetics*, vol 44, no. 9, pp 1699-1710, Online: http://dx.doi.org/10.1109/TCYB.2014.2336375

Venrooij, J, van Paassen, MM, Mulder, M, Abbink, DA, Mulder, M, van der Helm, FCT & Bülthoff, HH (September 2014), "A Framework for Biodynamic Feedthrough Analysis – Part I: Theoretical foundations", *IEEE Transactions on Cybernetics*, vol 44, no. 9, pp 1686-1698, Online: http://dx.doi.org/10.1109/TCYB.2014.2311043

Venrooij, J, Abbink, DA, Mulder, M, van Paassen, MM, Mulder, M, van der Helm, FCT & Bülthoff, HH (July 2014), "A Biodynamic Feedthrough Model Based on Neuromuscular Principles", *IEEE Transactions on Cybernetics*, vol 44, no. 7, pp 1141-1154, Online: http://dx.doi.org/10.1109/TCYB.2013.2280028

Venrooij, J, Mulder, M, Abbink, DA, van Paassen, MM, Mulder, M, van der Helm, FCT & Bulthoff, HH (July 2014), "Mathematical Biodynamic Feedthrough Model Applied to Rotorcraft", *IEEE Transactions on Cybernetics*, vol 44, no. 7, pp 1025-1038, Online: http://dx.doi.org/10.1109/TCYB.2013.2279018

Venrooij, J, Pavel, MD, Mulder, M, van der Helm, FCT & Bülthoff, HH (December 2013), "A Practical Biodynamic Feedthrough Model for Helicopters", *CEAS Aeronautical Journal*, vol 4, no. 4, pp 421-432, Online: http://dx.doi.org/10.1007/s13272-013-0083-y

Quaranta, G, Masarati, P & **Venrooij, J** (May 2013), "Impact of Pilots' Biodynamic Feedthrough on Rotorcraft by Robust Stability" *Journal of Sound and Vibration* vol 332, no. 20, pp. 4948-4962, Online: http://dx.doi.org/10.1016/j.jsv.2013.04.020

Venrooij, J, Mulder, M, Abbink, DA, van Paassen, MM, van der Helm, FCT, Bülthoff, HH & Mulder, M (February 2013), "A New View on Biodynamic Feedthrough Analysis: Unifying the Effects on Forces and Positions", *IEEE Transactions on Cybernetics*, vol 43, no. 1, pp. 129-142, Online: http://dx.doi.org/10.1109/TSMCB.2012.2200972

Venrooij, J, Abbink, DA, Mulder, M, van Paassen, MM & Mulder, M (August 2011), "A Method to Measure the Relationship Between Biodynamic Feedthrough and Neuromuscular Admittance", *IEEE Transactions on Systems, Man, and Cybernetics, Part B: Cybernetics*, vol 41, no. 4, pp. 1158-1169, Online: http://dx.doi.org/10.1109/TSMCB.2011.2112347

Conference publications

Venrooij, J, Mulder, M, Abbink, DA, van Paassen, MM, Mulder, M, van der Helm, FCT & Bülthoff, HH (October 2014), "Admittance-adaptive Model-based Cancellation of Biodynamic Feedthrough", *IEEE International Conference on Systems, Man, and Cybernetics, San Diego, USA, (SMC 2014)*, pp. 1961-1966, Online: http://dx.doi.org/10.1109/SMC.2014.6974206

Masarati, P, Quaranta, G, Zaichik, L, Yashin, Y, Desyatnik, P, Pavel, MD, **Venrooij, J** & Smaili, H (September 2013), "Biodynamic Pilot Modelling for Aeroelastic A/RPC" *39th European Rotorcraft Forum, Moscow, Russia (ERF 2013)*, pp. 1-14.

Venrooij, J, Mulder, M, van Paassen, MM, Abbink, DA, van der Helm, FCT, Mulder, M & Bülthoff, HH (October 2012), "How Effective Is an Armrest in Mitigating Biodynamic Feedthrough?", *IEEE International Conference on Systems, Man, and Cybernetics, Seoul, Korea, (SMC 2012)*, pp. 2150-2155,
Online: http://dx.doi.org/10.1109/ICSMC.2012.6378058

Venrooij, J, Pavel, MD, Mulder, M, van der Helm, FCT & Bülthoff, HH (September 2012), "A Practical Biodynamic Feedthrough Model for Helicopters", *38th European Rotorcraft Forum, Amsterdam, the Netherlands (ERF 2012)*, pp. 1-13

Quaranta, G, Masarati, P & **Venrooij, J** (May 2012), "Robust Stability Analysis: a Tool to Assess the Impact of Biodynamic Feedthrough on Rotorcraft" *68th American Helicopter Society International Annual Forum, Fort Worth (TX), USA, (AHS 2012)*, pp. 1306-1315.

Venrooij, J, Yilmaz, D, Pavel, MD, Quaranta, G, Jump, M & Mulder, M (September 2011), "Measuring Biodynamic Feedthrough in Helicopters", *37th European Rotorcraft Forum, Gallarate, Italy (ERF 2011)*, pp. 958-967

Venrooij, J, Mulder, M, van Paassen, MM, Abbink, DA, Bülthoff, HH & Mulder, M (October 2011), "Cancelling Biodynamic Feedthrough Requires a Subject and Task Dependent Approach", *IEEE International Conference on Systems, Man, and Cybernetics, Anchorage (AK), USA, (SMC 2011)*, pp. 1670-1675, Online:
http://dx.doi.org/10.1109/ICSMC.2011.6083911

Venrooij, J, Abbink, DA, Mulder, M, van Paassen, MM & Mulder, M (October 2010), "Biodynamic Feedthrough is Task Dependent", *IEEE International Conference on Systems, Man and Cybernetics, Istanbul, Turkey, (SMC 2010)*, pp. 2571-2578,
Online: http://dx.doi.org/10.1109/ICSMC.2010.5641915

Venrooij, J, Mulder, M, van Paassen, MM, Abbink, DA & Mulder, M (September 2010), "A Review of Biodynamic Feedthrough Mitigation Techniques", *11th IFAC/IFIP/IFORS/IEA Symposium on Analysis, Design, and Evaluation of Human-Machine Systems, Valenciennes, France*, pp. 316-321, Online: http://www.ifac-papersonline.net/detailed/47121.html

Venrooij, J, Abbink, DA , Mulder, M , van Paassen, MM & Mulder, M (April 2010), "Understanding the Role of the Neuromuscular Dynamics in Biodynamic Feedthrough Problems" *6th Pegasus - AIAA Student Conference 2010*, pp. 1-11.

Venrooij, J, Mulder, M, van Paassen, MM, Abbink, DA & Mulder, M (October 2009), "Relating Biodynamic Feedthrough to Neuromuscular Admittance", *IEEE International Conference on Systems, Man and Cybernetics, San Antonio (TX), USA, (SMC 2009)*, pp. 1668-1673, Online: http://dx.doi.org/10.1109/ICSMC.2009.5346935